Joseph William Williams, William Denison Roebuck, John W Taylor

Land and Fresh-Water Shells

An Introduction to the Study of Conchology. Second Edition

Joseph William Williams, William Denison Roebuck, John W Taylor

Land and Fresh-Water Shells
An Introduction to the Study of Conchology. Second Edition

ISBN/EAN: 9783337139759

Printed in Europe, USA, Canada, Australia, Japan

Cover: Foto ©berggeist007 / pixelio.de

More available books at **www.hansebooks.com**

Fig. 1.—A Group of Land Snails. [*Frontispiece.*

Helix arbustorum. *Helix fructicum.*
Helix hortensis.
Helix nemoralis.

Young Collector Series.]

LAND AND FRESH-WATER SHELLS:

AN INTRODUCTION TO THE STUDY OF CONCHOLOGY.

BY

J. W. WILLIAMS, M.A., D.Sc., F.S.L.A.,

GOLD MEDALLIST, S.Sc., LONDON;

Member of the Conchological Society of Great Britain and Ireland; Member of the Scottish Meteorological Society; Conchological Referee to the Wesley Scientific Society; Author of "The Shell-Collector's Handbook for the Field."

WITH A CHAPTER ON THE

DISTRIBUTION OF THE BRITISH LAND AND FRESH-WATER MOLLUSCA,

BY

J. W. TAYLOR, F.L.S.,

Editor of "The Journal of Conchology"; Ex-President of the Conchological Society of Great Britain and Ireland; and Membre Honoraire de la Société Malacologique de France;

AND

W. DENISON ROEBUCK, F.L.S.

SECOND EDITION

LONDON:
SWAN SONNENSCHEIN & CO.,
PATERNOSTER SQUARE.

1892.

Inscribed

TO THE MEMORY OF THE

GREATEST CONCHOLOGIST OF HIS TIME,

J. P. R. DRAPARNAUD

(BORN 1772, DIED 1804).

CONTENTS.

CHAPTER I.
                                                                                              PAGE

COLLECTING AND PRESERVING SNAILS, MUSSELS, AND SLUGS   9

CHAPTER II.
THE ANATOMY AND PHYSIOLOGY OF A SNAIL   14

CHAPTER III.
THE ANATOMY AND PHYSIOLOGY OF A FRESH-WATER MUSSEL   45

CHAPTER IV.
THE CLASSES, ORDERS, FAMILIES, GENERA, SPECIES, AND VARIETIES OF BRITISH LAND AND FRESH-WATER SHELLS   59

CHAPTER V.
THE DISTRIBUTION OF THE BRITISH LAND AND FRESH-WATER MOLLUSCA (J. W. TAYLOR AND W. DENISON ROEBUCK)   94

PREFACE.

THE burden of this little volume of "The Young Collector Series" is to supply a want often felt and often expressed by our younger naturalists, who cannot afford to give the high prices asked for books upon the subject of which it treats. At the same time it must not be looked upon as a finished monograph, nor yet as the only manual that will be needed by the student. As he makes advances in the after-days he will require the larger text-books of Moquin-Tandon, Gwyn Jeffreys, Forbes and Hanley, Rossmässler, and my other work, "The Shell-Collector's Handbook for the Field," to which this little volume may serve as an introduction. In the following text, the anatomy of the snail and fresh-water mussel has been given at some length, since the basis of our work must be founded on structural details, and to make these chapters as full as possible, all the available sources of Continental and Home literature have been searched for information. The whole of the species of our land and fresh-water mollusca are described, and their most common varieties are given in footnotes. To my friends, Mr. J. W. Taylor, F.L.S., and Mr. W. Denison Roebuck, F.L.S., I must express my indebtedness for allowing me to reprint here the chapter on distribution from *The Journal of Conchology*, which they have kindly brought up to date for the purposes of this volume; and I must also express the obligations I owe to that great army of workers (too numerous to mention individually) who, of late years, have done so much towards a fuller understanding of the subject, and from whose published works and papers I have largely culled much interesting and valuable information. My last word, in conclusion: should the young collector be ever at a loss to name a snail or a slug, let him send the specimen to me, enclosing the return postage, when I will do all in my power to help him over his difficulty.

> "Go, child of nature, to thy mother's breast,
> And learn the lesson she can teach so well;
> No longer in the lap of folly rest,
> But hear the truths that nature loves to tell."

J. W. W.

MIDDLESEX HOSPITAL, LONDON, W.,
AND MITTON, STOURPORT, WORCESTERSHIRE.

LAND AND FRESH-WATER SHELLS.

CHAPTER I.

COLLECTING AND PRESERVING SNAILS, MUSSELS, AND SLUGS.

IN my "Shell-Collector's Handbook for the Field," I have laid down five rules to be strictly followed when out collecting, which, advisably, may be recapitulated here. They are as follows:—

1. Never leave a stone unturned.
2. Never pass by a nettle without a full examination of its stem, branches, leaves, and the vegetation which grows below and around it.
3. Never leave untouched, and unexamined, moss at the roots of trees, or the dead leaves under them.
4. Always examine the vegetation on walls, and the grass which grows around their foundations.
5. Never forget, when searching for water-specimens, to examine the water-plants, and the under surface of any floating log of wood, as well as the bed of the pond, brook, or stream. An examination of caddis-cases will also often give a large number of water specimens.

In a review of the book from whence the above rules have been extracted, which appeared in one of our Natural History "Monthlies," the reviewer in touching upon these points expressed himself to the effect that when an individual is observed to be faithfully carrying them out "we may rest assured that we are watching the shell-collector on the war-path." No; "the shell-collector" is only "on the war-path" when he searches *everywhere*, for snails and slugs are omnipresent creatures, and are not, on the whole, very particular as to where they make their abode. The rules are what I term "golden rules;" *i.e.*, rules not to be forgotten; but the one rule of the assiduous collector is to search in every crink and cranny, and never to pass over a spot without a full examination of its *whole* surface. The implements needed for this collecting are few in number and inexpensive. Two of them are given you by nature—nimble fingers and quick eyes; one forms a companion in your walks,—the walking-stick; the others, which will have to be purchased, are a dredge, a scoop, or a water-net, and a few chip boxes. The walking stick will be useful for pulling water-weeds to the bankside, so that they may be easily examined with the fingers for specimens. The dredge will be needed for the collection of *Unios, Anodons, Sphæria*, and *Pisidia*, which, being bivalves, live almost entirely on the bed of the pond or stream. This can be made after the following figure of a net-dredge recommended in Messrs. Gray

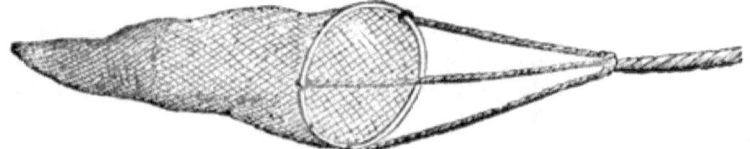

Fig. 2.—A Water-dredge (Woodward).

and Woodward's companion volume, "Seaweeds, Shells, and Fossils;" but my preference is to having a properly weighted square frame rather than a circular one, as it takes a larger area of the water-bed at the same time, and, consequently, is more apt to work up a larger number of specimens. The scoop is not in general use, I think, but, nevertheless, it is an exceedingly good implement to use for sweeping among water-weeds which cannot be brought to the bankside with the stick, and perhaps is more welcome and more serviceable than the ordinary water-net, for which it forms a substitute. Mr. Wallis Kew and I have taken a large number of specimens with it from the Tottenham Marshes, —by the way, a happy hunting ground for the London collector. Mr. Denison Roebuck has his made of copper, but zinc will be found cheaper and nearly as serviceable a material. A circular ring of zinc is made, of about an inch and a quarter in depth, and of about eight inches in diameter. A perforated zinc bottom is then soldered on, and also a zinc handle for a stick to fit into is made and soldered on, or better, riveted to the zinc ring, so that, when finished, it resembles in shape, more or less, a small brewing sieve. Thus it is stronger than the net, and can be wielded with a more powerful hand without much fear of mishaps occurring to spoil the day's collecting. Should, however, the net be preferred, it should be made of strong book-muslin, of an oblong bag-shape, and the ring should be of thick galvanised iron. The chip boxes can be bought in nests from any dealer at a cheap rate, and their utility consists simply in forming pocketable boxes in which to carry home your specimens. The paraphernalia, then, is only wanted for water-specimens; for land-shells and slugs nothing is needed but your own fingers and eyes.

The specimens, we must now assume, are collected, and taken home to be made ready for a place in the collection. How are they to be made ready? In the first place, they must be killed; in the case of the snails, by boiling water, and, in the case of the slugs, by drowning. The animals of the shells must then be extracted. With the univalves this is done best with a bent pin; with the bivalves the animal is best extracted by the small blade of the pocket-knife, or by a small scalpel. The shells of the univalves are then well cleaned by hard brushing with a tooth-brush dipped in pure water; the shells of the bivalves are to have their two valves tied together tightly with cotton or string so that the ligaments may dry, and so prevent them from divaricating, after which they are to be cleaned in the fashion employed with the univalves. Mr. Roebuck is not inclined to clean his shells, as he thinks the amount of confervoid growth often interesting; but this must be left to the personal ideas of the collector alone.

It is quite another method with the slugs. Their shells may be extracted from underneath the mantle (except in *Testacella*, where the shell is situated on the tip of the tail) by the pen-knife, and glued on cardboard or kept loose in chip boxes, or the animal itself may be preserved in one of two ways. It may be preserved by the dry or by the wet method. There are two dry methods. In one—recommended by Hübner—the animal is killed by immersion in methylated spirit, and the whole of the viscera is extracted through a longitudinal incision made along the under surface of the foot in the middle line; the place which the viscera held is stuffed with wadding, and the skin is dried and varnished. In the other—recommended by Dubreuil—the slug is killed and washed in pure water, to which, after the lapse of eight hours, some salt is added. A slit is then made along the left side, and the animal skinned. Thus, by means of two more longitudinal slits, three preparations can be made—one to show the back, one the foot, and the third the right side with the pulmonary orifice. These are glued on cardboard, varnished with white shellac varnish, to which a little corrosive sublimate has been added, and duly labelled.*

The wet method is simply their immersion in some one fluid preservative medium. This may be turpentine, equal parts of glycerine and methylated spirit, or preferably one of the following fluids, which can be made with slight trouble.

(1). *Wickerheimer's Fluid.*

Dissolve in 3,000 grammes of boiling water—

100	,,	alum,
25	,,	common salt,
12	,,	saltpetre,
60	,,	carbonate of potash,
10	,,	arsenious acid.

This solution is to be cooled and filtered, and to every ten litres of it four litres of glycerine and one litre of methylated spirit are to be added.

(2). *Poetz and Hohr's Fluid.*

Arsenious acid ...	...	12 grammes
Sodium chloride	...	60 ,,
Potassium sulphite	...	150 ,,
,, nitrate	...	18 ,,
,, carbonate	...	15 ,,
Water ...	...	10 litres
Glycerine ...	...	4 ,,
Wood-naphtha ...	...	¾ ,,

* There is a separate method for the preservation of *Testacella*. The animal is dried in sand, then slit along its under surface, stuffed with cotton wool and redried.

(3).

| Water | ... | ... | ... | ... | 800 parts |
| Chloride of zinc | ... | ... | ... | ... | 100 ,, |

Dissolve.

(4).

White sugar	...	...	...	...	100 parts
Methyl-spirit	...	...	...	...	400 ,,
Common salt	...	...	...	...	200 ,,
Nitrate of potash	...	...	...	...	50 ,,
Water	...	...	...	...	750 ,,

(5).

Hyposulphite of soda ... ... q. s. to make a saturated solution in

Water
Methylated spirit } equal parts.

The slugs, well cleared of their mucus, are to be placed in any one of the above media, in small cylindrical glass tubes and tightly sealed, so as not to admit air. This sealing can be done by means of tightly fitting corks, subsequently covered over with a coating of Brunswick black, or by the use of glass tops cemented down with a mixture of *old* guttapercha, five parts, and asphalt, four parts, melted together and applied hot.

In concluding this chapter, I would strongly advise you not to go to the expense of buying cabinets wherein to lodge your shells. Rather purchase large books on the subject with your money. Cardboard trays are all that is needed for storage purposes. In them lay the shells on cotton-wool, with a label bearing the date of capture and the locality, also the specific and—if a variety—the varietal name, but these are not so important as the locality, for the shells carry their own identification with them.

CHAPTER II.

THE ANATOMY AND PHYSIOLOGY OF A SNAIL.

TAKE a live Wrinkled Snail (*Helix aspersa*) or Roman Snail (*Helix pomatia*), and allow it to crawl along a piece of paper before you. Note—

1. That part only of the animal is extruded (*prostoma*), the other portion (*metastoma*, or *visceral hump*) being permanently lodged within the shell.

2. On the prostoma note—

(*a*) The *head segment*, forming its anterior extremity, and bearing two pairs of processes or tentacles; the upper pair, carrying the eyes (*dorsal tentacles*, or *ommatophores*), are much longer than the lower pair (*ventral tentacles*).

(*b*) The *mouth* on the anterior extremity of the head segment, bounded by a *circular* and a *lateral lip*.

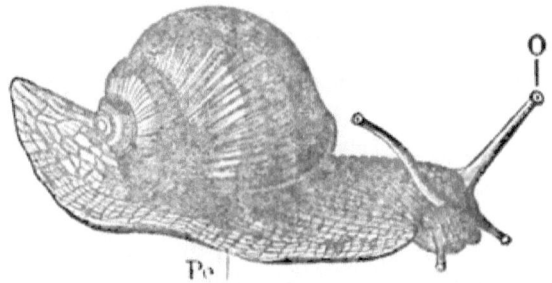

Fig. 3.—*Helix pomatia*. O, ommatophore; Pe, foot.

(*c*) The *foot*, forming a well-marked muscular expansion on the under surface of the prostoma.

(*d*) The *orifice of the pedal gland*, between the mouth and the foot.

(*e*) The *genital orifice*, slightly below and slightly behind the right dorsal tentacle.

(*f*) The *collar of the mantle*, a portion of the mantle-edge just visible beneath the aperture of the shell. Notice in this, on the right side, the *pulmonary aperture*, the *anus*, and the *external orifice of the duct of the renal gland*.

(*g*) The *pulmonary orifice*, a large foramen situated on the right side of the mantle-collar, and leading into the mantle cavity.

(*h*) The *anus*, opening side by side with, and slightly to the right of, the pulmonary orifice.

(*i*) The *orifice of the duct of the renal gland*, opening just inside the pulmonary orifice.

Kill the snail by drowning, and place in spirit for a few hours. Twist the animal out of the shell, and replace it in the spirit. On the shell note—

(*a*) The colour and markings.

(*b*) Its conical shape. This may be discoid, as in *Planorbis complanatus*; oval, as in *Amalia marginata*; or ear-shaped, as in *Testacella haliotidea*.

(*c*) Its division into four *whorls*, the largest being known as the *body-whorl*.

(*d*) The *aperture, peristome*, or *peritreme*, through which the snail extends its prostoma. This is toothed in some species, as *Pupa secale, Vertigo pusilla*, and *Vertigo antivertigo*.

(*e*) The *nucleus*, or apex of the cone, the oldest part of the shell corresponding to the umbo of the Lamellibranch shell.

(*f*) The *lines of growth*, running parallel to the aperture.

(*g*) The *varices*, numerous well-marked lines extending across the body-whorl, and differentiating one year's growth from another.

(*h*) The *columella*, or axis around which the whorls turn.

(*i*) The *umbilicus*, a foramen on the under surface of the shell, opening into the cavity of the columella.

(*j*) In a hibernating specimen, notice the *epiphragm*, which completely closes the mouth of the shell. Examine with a lens its under surface, and notice a small opening in its central portion for the admission of air for respiratory purposes.

THE SHELL GENERALLY CONSIDERED.—The shell of the adult Gastropod, with the exception of the internal shell of *Arion* and *Limax* and the nucleus of *Clausilia*, is a *secondary* growth formed by the mantle, not by the shell-gland of the embryo. In structure, according to the researches of Longe and Mer, it may be said to consist of three layers,—a *cuticle* (*epidermis, periostracum*), a *honeycomb layer*, and a *nacreous* or *mother-of-pearl layer*. The first of these layers is formed by a cutogenic apparatus composed of an "*epithelial organ*" of bottle-shaped cells secreting granules, and of numerous glandular cæca which open into a line—the "*pallial line*"—situated just behind the collar of the mantle. Nalepa has found that in *Zonites algirus* the cells of this "epithelial organ" develop in spring, the period of greatest growth, but after that—*i.e.*, during summer, autumn- and winter—they gradually atrophy, and, according to Longe and Mer, they are entirely wanting in the full-grown animal. The colour of the shell is secreted by the unicellular glands of the mantle, the second layer by the edge of the mantle, and the third layer by the whole surface of the mantle. The honeycomb layer is composed of a superficial and a deeper portion; the former shows a confused striated appearance, the latter consists of vertical prisms. The nacreous layer is arranged in three super-posed series of calcareous lamellæ, those of the outer and inner layers running parallel to the suture, and the middle layer at right angles to it. Splinters of the shell are hard enough to scratch Calc-spar; the form in which the calcareous matter exists is represented by Arragonite ($CaCO_3$) in the mineral kingdom. In chemical composition the shell is made up of about 95-98 per cent. of carbonate of lime, of about 1·5 per cent. of an organic substance termed conchiolin, and of magnesium carbonate, silica, and alumina, with phosphates, in small quantities.

The mouth of the shell is closed in some species by an *epiphragm*, in others by an *operculum*, and in others still by a *clausilium*. Each of these structures requires a few words. The epiphragm is a film of mucus impregnated by calcareous salts secreted over the aperture of the shell in some species, as the *Helices*, previous to hibernation. Barfurth has shown that the calcareous salts are a secretion of the cells of the digestive gland. The adhesion that takes place between the epiphragm and the peristome of the shell is dissolved away by an excretion of fluid mucus from the mantle, at what time the trees under the influence of returning warmth first burgeon into shoots of freshest green, after which it is easily thrown off by the pressure of the foot. Hammersten finds that the foot of some snails, as *H. pomatia*, excretes mucus as well as the mantle, and possibly this excretion also helps in the disintegration of the epiphragm. The name of epiphragm was first given to this structure by J. P. R. Draparnaud, to whose memory this present little volume of conchology is inscribed; old Martin Lister (born 1638 (?), died 1712), naturalist and court physician, who thought he had done everything in natural history that could be done, called it the operculum saliva confectum; Müller termed it the operculum hybernum. The operculum proper is a horny or shelly plate attached to the back of the foot in many water-snails, as *Paludina*, *Neritina*, *Bythinia*, and *Valvata*, and in a few land snails, as *Cyclostoma* and *Acme*. It contains more conchiolin in its composition than the shell whose mouth it closes, and not less conchiolin and more chitine, as Dr. Henry Woodward has recently supposed. Those Gastropods which have no operculum are spoken of collectively as the *inopercular univalves;* those with an operculum, on the other hand, as the *operculated univalves*. No operculum of any shell, whether of fresh-water, land, or marine habitat, exhibits an annular form. In a marine form, *Lithedaphus* (*Calyptræa*) *equestris*, an operculum is attached to the whole length of the foot, so that on first appearance the shell might be taken to be a bivalve. But as Professor Owen has remarked that upon a comparative study of the operculum it gives characters of secondary importance since it " sometimes varies in structure in species of the same genus, as it is present in some *Volutes*, *Cones*, *Mitres*, and *Olives* (marine forms), and absent in other species of those genera, and as some genera in a natural family, as *Harpæ* and *Dolium* among the Buccinoids, are without an operculum, whilst the other genera of the same family possess that appendage." The operculum grows by the addition of matter to its circumference, its youngest part being called the *nucleus*, which may be excentric, as in *Paludina vivipara*, or central, as in *Bythinia Leachii*. The clausilium is characteristic of the genus *Clausilia*, and differs from the operculum in not being attached to the animal. It may be best seen by taking a *Clausilia laminata* and breaking away the outer part of the body-whorl, when it will be found to consist of a shelly plate attached to the columella by an elastic ligament about half a whorl from the aperture. It serves exactly the same function in the economy of the animal as an operculum does among the Paludinidæ and Cyclostomatidæ. When the animal extends itself out of the shell the clausilium is pushed back against the columella, and when it withdraws, the clausilium flies backwards on account of the elasticity of its ligament, and closes the mouth of the shell.

The *aperture of the shell*, generally spoken of as the *mouth peristome*, or *peritreme*, varies greatly in shape in different species. When it is entire the shell is called *holostomatous;* when produced into a canal for a siphon it is spoken of as *siphonostomatous*. It is sometimes reflected over the umbilicus; this is due to the presence of a lobe in the collar of the mantle of the animal, termed the *columellar lobule*. It was John Hunter, the great anatomist, (born 1728, died 1793), who first noticed that the animal has the power of absorbing part of its shell, and this will explain the thinning of some parts of a whorl in a shell in relation to the other portions, as is sometimes observed. In some species, as *Bulimus decollatus*, the apex becomes, as the animal grows to the adult condition, *decollated ; i.e.*, the physical decomposition of the apex of a shell, due to the animal leaving it for the lower and larger whorls as it grows in size. When a shell is injured at the peristome all the layers are reproduced; when at any other portion only the middle and internal layers are repaired, and then the shell is generally thickened internally.

The spire of the majority of the molluscan shells turns to the right, and the mouth, when placed in its proper position, looks towards the left; it is then spoken of as *dextral* or *dexiotrope*. In some genera, as *Physa*, *Balia*, and *Clausilia*, the reverse of this obtains, and then the shell is said to be *sinistral* or *laiotrope*— the spire turns towards the left, and the mouth, placed in its proper position, looks towards the right.

The colour of the shell varies in different species, and often in the same species. Camerano has studied the comparative rarity and frequency of the colours of the shells of the Mollusca, and his inferences are that black is rare; brown, grey, yellow, white, and red common; violet relatively abundant; blue not rare; and green infrequent. Some shells are unicolorous and not banded; others are unicolorous or banded. Mr. J. W. Taylor, in his recent "Valedictory Address as President to the Conchological Society for the year 1887," published in *The Journal of Conchology* for April, 1888, makes the following statement: "Nearly seven years ago Mr. Ashford suggested to me the probability that *Helix cantiana, cartusiana*, etc., were once banded species, and I am disposed to agree with his suggestion." But there seems to the writer scarcely any foundation for such a supposition, and he would disagree with such a statement for five reasons. In the first place, the primary shell in the embryo of all species is unicolorous; in the second place, the youngest portion (nucleus) of the secondary shell in all species is unicolorous; in the third place, unicolorous specimens are most common in aquatic forms, where environmental circumstances have doubtless not been so great as on land; in the fourth place, of a banded species unicolorous specimens are invariably found; and in the fifth place, such a supposition is not general and comparative enough. Thus, the writer is inclined to the opinion that all the species of our land and fresh-water mollusca were once, and only, unicolorous. What the *causa causans* of the future banding was, we are, practically at any rate, in doubt. It remains as one of Nature's greatest arcana for the great future to solve.

In *H. nemoralis* and *H. hortensis* the bands vary in number and degree of development, so that it is necessary to have some system by which we can represent these variations on paper for the use of our fellow-workers. The system is known by the name of the *system of the band-formulæ*. The type of each species has five bands, which are represented by the figures 12345, the 123 referring to three bands above the periphery, and the 45 to two bands below it. When a band is interrupted it is shown by a colon, thus, 12:45; when slightly developed, by the insertion of the figure representing that band below the line of the others as 12₃45, when obsolete by a cypher, as 02345, or 00000; and when two or more bands are fused together into one band it is shown by bracketing the figures standing for the component bands, thus, 123 (45), (123) (45). If more than five bands be present the extra band is represented by the insertion of the letter x into the formula, as 12 × 345 or 1 × 2345.

Sometimes a shell becomes shaped like a spiral ladder, or *scalarid*, as in the accompanying figure of the *m. scalariforme* of our common Wrinkled Snail. The shell in *Arion* may be absent,

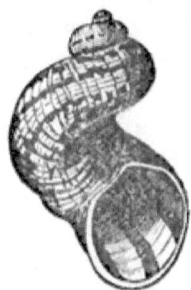

Fig. 4.—*Helix aspersa.* m. *scalariforme.*

or may consist of a few calcareous granules: in *Limax*, the shell is shield-like; in both it is situated underneath the mantle. The grades may be traced from these simple shells to the more developed shells of the Helices, through the little ear-shaped shell on the tail of *Testacella haliotidea*, and the pretty glassy-green shell of *Vitrina pellucida*.

THE BODY-WALL CONSIDERED GENERALLY.—In the *Limacidæ* the mantle is visible throughout its whole extent; in *Limax* the respiratory orifice is on its posterior half; in *Arion* and *Geomalacus* on its anterior half. It may partly overlap the shell, as in

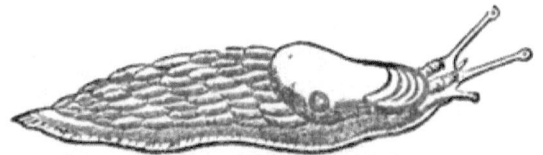

Fig. 5.—A slug (*Limax*) showing the *respiratory orifice*.

Limnæa glutinosa, and in *Physa fontinalis*, or its fore-edge may undergo concrescence with the dorsum of the body-wall, as in the *Helicidæ*. In the *Limnæa* the fore-edge is free, and no such concrescence has taken place. In *Valvata* the long plume-like gills protrude from the mantle cavity when the animal is walking.

There may be four cylindrical tentacles, as in the *Limacidæ* and *Helicidæ*, or only two, as in *Limnæa* and in *Physa*. When two tentacles only are present they may be short and triangular, as in *Limnæa*, very long and cylindrical, as in *Planorbis* and *Physa*. The eyes may be situated on the apices of the dorsal tentacles, as in *Helix*, or situated at their base, as in *Limnæa*, *Ancylus*, and *Physa*, or sessile, as in *Planorbis*, or placed upon tubercles, as in *Hydrobia*, or placed upon pedicels, as in *Paludina*.

The reproductive orifice is generally a single opening just behind and slightly below the right tentacle, but in some genera, as *Limnæa*, *Planorbis*, and *Ancylus*, the generative openings are separated from one another, the male orifice (*penial orifice*) being situated a little anterior to the female orifice (*os vaginæ*), a condition of things which seems to be associated with the situation of the eyes at the base of the tentacles. To those genera in which these two last-mentioned features obtain Professor Ray Lankester has given the name of BASOMMATOPHORA, as distinguished from the STYLOMMATOPHORA, in which there is only one generative aperture, and the eyes are elevated on tentacles which can be invaginated into the body-cavity, as the genera *Helix*, *Pupa*, *Clausilia*, *Testacella*, *Limax*, *Amalia*, *Arion*, and *Onchidium*.

The visceral mass may be large and coiled spirally in the shell, as in the *Helicidæ*, *Limnæidæ*, *Cyclostomatidæ*, *Paludinidæ*, and *Carychiidæ*, or it may be co-extensive with the foot, as in the *Limacidæ*. The sole of the foot is ciliated in the majority of terrestrial forms, and this ciliation extends all over the surface of the body in the aquatic species. In *Arion* the lateral margins of the foot are also ciliated, and this feature also obtains around the margin of the supra-pedal gland and the pulmonary orifice in *Helix nemoralis* and in *Amalia marginata*. In microscopical structure the body-wall consists of an *epidermis* of columnar cells, and internally to this a thickened layer of muscle-fibres and connective tissue. The muscle-fibres are not so differentiated as those in the muscle-tissue of vertebrates, but consist of cells which generally are non-striated, although sometimes they are found to present an appearance of fine striation. The connective tissue is described as made up of " plasma-cells, a matrix, and fibrils." The plasma-cells may be irregular in shape, containing refractile granules, or oval, nucleated, with a clear protoplasm, or they may contain granules of carbonate of lime. In the oval-shaped cells

Blundstone has found a substance named glycogen in *Helix*, and also in the fresh-water mussel (*Anodon*), an observation to which we shall have to refer when treating about the physiology of the snail. The matrix contains stellate-shaped cells, and in *Amalia carinata* and *Limax flavus* Leydig has observed coloured cells, termed *chromoblasts*, which change their contour under the influence of suitable re-agents. Mucous glands, unicellular in character, are scattered all over the body of all Pulmonates, together with goblet-shaped cells and sensory cells. The mucous glands are in great quantity on the collar of the mantle; they secrete mucus, and it is this mucus which forms the threads by which the slugs and the *Limnæa*, among land and fresh-water molluscs, and the *Litiopa* and *Rissoa parva*, among marine forms, suspend themselves.

Running along the foot in all Pulmonates is a cæcal diverticulum of the integument termed the supra-pedal gland. In *Cyclostoma elegans* there are two. In *Arion* and *Geomalacus* a large slime gland is present at the posterior extremity of the tail; in *Amalia* and *Limax* this structural feature is absent. The epithelium of the supra-pedal gland is ciliated, and has been said to contain sense cells.

Pin the animal down under water, and reflect the mantle and the integument of the prostoma. Dissect the following systems out in order:—

THE DIGESTIVE SYSTEM.—The digestive system commences at the mouth, already noticed, enlarges to form a buccal mass, then passes back by an œsophagus to form a crop and stomach, and from thence proceeds through the substance of the digestive gland, or " Mitteldarmdrüse," as Frenzel terms it, as the intestine to end at the anus. The *buccal mass* or *pharynx* is a stout muscular cavity, situated in the head segment, from the lower and posterior surface of which a pale diverticulum depends,—the *sac of the radula*. This buccal mass contains the masticatory organs. These consist of a *horny beak* lying in the circular lip, and of an *odontophore* lying below a yellow ribbon containing teeth, known as the *radula*.

The beak plays the function of a mandible in grasping the vegetable or animal substance on which the animal feeds, while the radula with its numerous teeth rasps it off into the buccal cavity. In some species, as *Testacella*, there is no beak at all, in others it is either single, double, or treble. It also differs in contour, and, according to these differences in shape or number, naturalists

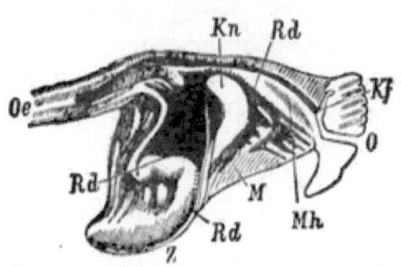

Fig. 6.—Longitudinal section through the Buccal Mass of *Helix* (after W. Keferstein). *O*, mouth; *Mh*, buccal cavity; *M*, muscles; *Rd*, radula; *Kn*, lingual cartilage; *Œ*, œsophagus; *Kf*, jaws; *Z*, sheath of radula.

have striven to form a system of classification, and none, perhaps, more so than Moquin-Tandon and Dr. Mörch. It has not, however, found as much favour with English workers as it deserves, but on the continent it is almost universally adopted by conchologists, and is also, now and again, to be found referred to in English works and papers on conchology. The student must, therefore, make himself perfectly familiar with the terminology of this portion of the subject, and the following synopsis is adopted from an article communicated by Mr. Ralph Tate to the *Intellectual Observer* about twenty years ago, but which contains, nevertheless, a full explanation of the terms necessary for the beginner to make himself conversant with.

Synopsis of the genera of British Land and Fresh-water Mollusca, grouped according to the nature of the mandibular organ.

I.—AGNATHA (without mandibles). *Testacella.*

II.—MONOGNATHA (mandible single).

    1. *Oxygnatha.* Mandible lunate, smooth, rostrated. *Limax, Amalia, Vitrina, Hyalina.*

    2. *Elasmognatha.* Horse-shoe-shaped, with a plate behind. *Succinea.*

    3. *Aulacognatha.* Sides nearly parallel, striated. *Bulimus, Cochlicopa, Clausilia, Balia, Pupa, Vertigo,*

Helix (*H. rupestris*, *H. pygmea*, *H. rotundata*, *H. pulchella*), Carychium, Physa, Planorbis (excluding *P. corneus*).

4. *Odontognatha*. Ribbed, margin denticulated or crenulated.

(*a*) Slightly arched, ribs numerous, bluntly denticulate margins. Arion, Geomalacus, Helix (*H. obvoluta*, *H. cantiana*, *H. cartusiana*, *H. sericea*, *H. hispida*, *H. concinna*, *H. rufescens*, *H. acuta*).

(*b*) Strongly arched, ribs few. Helix (*H. ericetorum*, *H. pisana*, *H. caperata*, *H. virgata*, *H. nemoralis*, *H. hortensis*, *H. aspersa*, *H. pomatia*).

III. TRIGNATHA. Mandibles three. *Limnæa*, *Ancylus*, *Planorbis corneus*.

IV. DIGNATHA. Mandibles two, lateral. *Cyclostoma*, *Acme*, *Paludina*, and the other fresh-water Gastropods.

The radula consists of a chitinous plate beset with teeth, which as they are worn away are replaced by others from behind. Taking a transverse row of teeth, it will be found that they vary in different places, and it is possible to represent them by using a formula. "Thus, when in each row there is a single median tooth

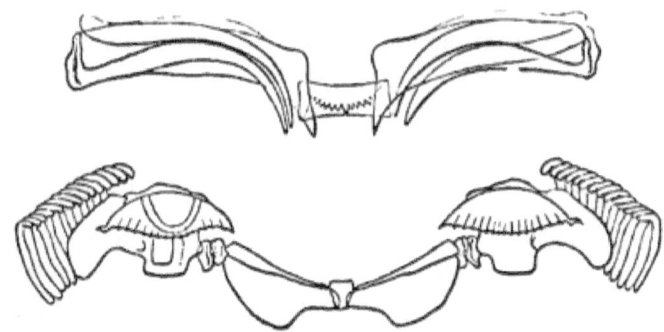

Fig. 7.—A segment of the radula of *Neritina fluviatilis* (after S. Lovén).

with three teeth on each side of it, more or less closely resembling one another, we write the formula 3.1.3. When there are additional lateral pieces of a different shape to those immediately adjoining the central teeth, we indicate them by the figure o, repeated to represent their number. Thus oooo.1.1.1.oooo is the formula for the lingual teeth of *Chiton Stelleri*. A single median tooth, an admedian series, and a lateral series may be thus distinguished. In some Glossophora * only median teeth are present, or large median teeth with a single small admedian tooth on each side of it; these are termed Rachiglossa (formula, —1.—or 1.1.1.). In a large number of Glossophora we have three admedian on each side and one median, no lateral pieces; these are termed Tænioglossa (formula 3.1.3.). Those with numerous lateral pieces, four to six or more admedian pieces, and a median piece or tooth, are termed Rhipidoglossa (formula x.6.1.6.x., where x. stands for an independent number of lateral pieces). The Toxoglossa have 1.0.1., the central tooth being absent, and the lateral teeth peculiarly long and connected with muscles. The term Ptenoglossa is applied to those Glossophora in which the radula presents no median tooth, but an indefinite and large number of admedian teeth, giving the formula x.o.x. When the admedian teeth are indefinite (forty to fifty) and a median tooth is present, the term Myriaglossa is applied (formula x.1.x). It must be understood that the pieces or teeth thus formulated may themselves vary much in form, being either flat plates, or denticulated, hooked, or spine-like bodies" (Professor Ray. Lankester, Art. "Mollusca." *Ency. Brit.*, vol. ix., pp. 640, 641). In some of the Helices the total number of teeth present on one radula may amount to more than thirty-nine thousand. The radula is supported by a subradular membrane, under which there are a pair of odontophoral cartilages to which intrinsic muscles are attached. The buccal mass is moved by numerous muscles. The *retractor muscle* is a large sheet which passes back from its floor to be inserted into the columella; the *protractors* are a number of delicate bundles of muscle tissue, which pass from its side walls to be inserted into the integument of the head; the *levators* are delicate like the last, and arise just above them to be inserted near the ventral tentacles; the *depressors* pass obliquely backwards, and are situated just below the protractors. MacMunn has found a colouring matter, termed myohæmatin, in these muscles, and Prof. Lankester has found that hæmoglobin is present in the

* Molluscs with an odontophore and radula.

buccal mass of some snails. The epithelium in the buccal mass is cylindrical, and in some places ciliated.

From the buccal mass a narrow gullet (*œsophagus*) passes into a distended portion of the enteric tract,—the *crop*,—on the sides of which are placed the *salivary glands*. A microscopical examination of a section of the œsophagus presents six layers from without inwards,—(1) a peritoneal investing coat; (2) a layer of longitudinally arranged muscle-cells; (3) a layer of circularly arranged muscle-cells; (4) a lacunar mucosa layer; (5) a layer of epithelium; (6) a lining cuticle. The crop is spindle-shaped (*fusiform*), with its lining membrane thrown into numerous longitudinal folds; it is generally conspicuous on account of the yellow character of its contents. The salivary glands are compound glands made up of unicellular glands. They are generally two in number, but *H. pomatia* has an additional one imbedded in its buccal mass. They secrete free sulphuric and free hydrochloric acids, which act upon the starchy matters of their food-stuffs in the same way as the ptyalin of the human saliva. They convert the starch into sugar. This secretion is carried from the glands by two ducts, which open into the buccal mass just above the odontophore. The glands will be seen to be lobulated, and confluent with one another on the posterior and dorsal surface of the crop.

Passing from the crop the enteric canal narrows, but soon expands again to form the *stomach*. This is a simple sac, incompletely subdivided into two by a longitudinal septum, and into which the ducts of the "*Mitteldarmdrüse*" open. The stomach differs in contour in different species. In some marine forms, as *Aplysia* and *Bullœa aperta*, horny processes and plates are developed, with which to triturate the food, a condition which strongly reminds one of the "gastric mill" of a cray-fish. In the former of these, *Aplysia*, the wall of the stomach is thickened, and beset with horny spines like canine teeth, and with rhomboidal-shaped plates like molar teeth. The food, consisting of seaweeds, after having been coarsely masticated in the buccal cavity and crop, is pierced by these spines and afterwards pounded into a pulpy mass by the plates. In the second of these, *Bullœa aperta*, there are three large plates, which are convex on the outside and concave interiorly. In a third case, *Bulla lignaria*, there are three plates, attached to which are strong muscles, but these plates are calcareous, and not horny, as are those of *Bullæa aperta*. It will be a reminder to those who are careless, to know that these calcareous plates in the stomach of *Bulla* have been wrongly described as a new bivalve shell, under the name of *Giænia*. Our

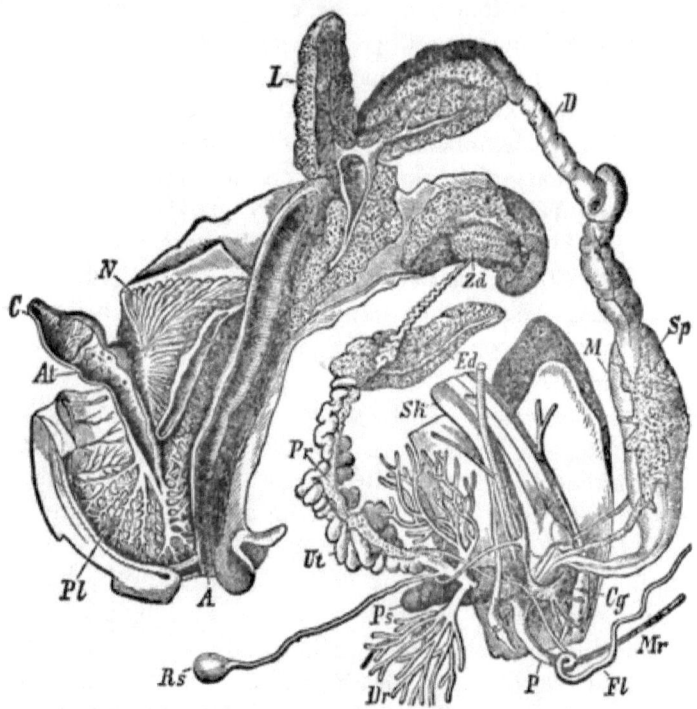

Fig. 8.—Anatomy of *H. Pomatia* (after Cuvier). The mantle-cavity is opened on the left side, and the mantle is turned over to the right. The body cavity has been opened, and the viscera are unravelled. *Cg*, supra-œsophageal ganglion; *Sp*, salivary gland; *M*, crop, *D*, intestine; *L*, "Mitteldarmdrüse" (digestive gland, hepato-pancreas, gland of mid-intestine, liver); *A*, anus; *At*, auricle; *C*, ventricle; *Pl*, lung; *Zd*, hermaphrodite gland invested by lobes of the "Mitteldarmdrüse"; *Ed*, albumi-niparous gland; *Pr*, prostate; *Ut*, female portion of the common generative duct; *Rs*, receptaculum seminis or spermatheca; *Dr*, muciparous glands; *Ps*, dart-sac; *P*, penis; *Fl*, flagellum; *Mr*, retractor muscle of the penis; *Sp*, spindle muscle.

own fresh-water form, *Limnæa stagnalis*, approaches the condition of things, I think, found in *Bullæa aperta*. It has a strong stomach, and is in the habit of swallowing pebbles from the bed of the pond, for the purpose, no doubt, of aiding in the trituration of its food. In one stomach of the animal of that species I counted the other day no less than twenty-seven small pebbles.

The "Mitteldarmdrüse," as the organ has been recently called by Frenzel, is a brownish mass divided into two asymmetrical portions, and lying in the whorls of the spire. Known in English works as the *liver, hepato-pancreas, digestive gland*, or *gland of the mid-intestine*, its right lobe or portion is the larger, and is subdivided into three smaller lobes, from each of which a duct carries the secretion to a common duct, which opens on the right side of the stomach. It is grooved by the intestine, and occupies the upper half of the first turn of the spire. The left portion, or lobe, separated from the right lobe by four structures,—the crop, stomach, intestine, and albuminiparous gland,—is placed in the upper turns of the spire, and has imbedded in its substance the *hermaphrodite gland* or *ovotestis*. It has a single duct, which opens into the left side of the stomach, nearly opposite to the opening of the duct of the right lobe. In microscopical structure it is an acinous gland, the several acini of which are bound together by connective tissue. The gland substance shows a differentiation into three kinds of cells,—*calcareous cells*, which are triangular, and contain lime in combination, according to Barfurth, with phosphoric acid;[*] *hepatic* or *granular cells*, containing fat globules, albuminous bodies, and granules which are highly refractile—the fat globules are passed with the fæces; *ferment cells*, which are more or less club-shaped, and contain fat globules, albumen globules, and other bodies that are more or less viscous. The fat globules blacken on treatment with osmic acid. The ferment cells secrete an unformed ferment named trypsin, and an amylolytic ferment. In the "Mitteldarmdrüse" of *Arion ater, Helix aspersa*, and *H. pomatia* hæmatin has been demonstrated; and in the last named of these species enterochlorophyll has been found in addition. According to Barfurth, the secretion of the calcareous cells goes in the main to the formation of the epiphragm. He burnt the organ to an ash at different periods of the year, and then quantitatively estimated the amount of lime it contained. The average amount present in May was 20·24 per cent., in September 25·72 per cent., and after the formation of the

[*] This has been opposed by Frenzel.

epiphragm 10·26 per cent. The liver secretion is acid to litmus paper. In *Planorbis, Limnæa,* and *Buccinum* the œsophagus has a lateral cæcal crop appended to it. The anus terminates on the left side of the body in *Planorbis,* and it has a median position in *Testacella* and *Doris.*

The trypsin of the secretion of the "Mitteldarmdrüse" converts the proteids of the foodstuffs into peptones, so that they can be absorbed by the enteric tract; the amylolytic ferment converts starch into sugar, and so renders it into a condition necessary for absorption. This absorption probably goes on throughout the whole length of the crop, stomach, and intestine, and more especially by the epithelium covering certain conical processes of the mucosa layer of those portions of the enteric tract. But the most interesting fact in the physiology of the Mollusca is that a substance termed glycogen has been found in most of their tissues, and especially in the substance of this digestive gland,—the "Mitteldarmdrüse." This glycogen ($C_6H_{10}O_5$), as can be seen from its formula, is an isomer of starch and dextrin. It is a white amorphous substance, insoluble in alcohol and ether, and its aqueous solution has a strong dextro-rotatory influence on polarised light. Its presence in the tissues may be readily detected—provided the animal has been recently killed—by testing with iodurretted potassium, with which it gives a wine-red colour, disappearing on heating, reappearing on cooling. In the Vertebrates, it has been detected in the liver (1·5—4 per cent.), muscles, villi of the chorion, embryonal tissues, placenta, and the white corpuscles of the blood. In the Mollusca, Hammersten has demonstrated that in the liver of *Helix pomatia* it is present to the extent of 1·75 per cent., and that in hibernating animals of the same species, and in the same organ, it is decreased to the amount of 0·429 per cent.; Barfurth has found it in most of the tissues of *Arion, Limax, Helix* and *Cyclostoma;* while Blundstone has discovered it in the mesentery of *Helix.*

Whence comes this glycogen, and whither is it bound? We may get a suggestion or two regarding these questions which we naturally ask, if we will call to our mind what is known of the physiology of the Vertebrates in this particular, and compare it with what has been demonstrated as existing in the Mollusca.

There can be little doubt that the mother-substance of glycogen in the Vertebrates, at any rate, is carbohydrate food, while the ingestion of proteid matter seems also to favour its production, but in a far less degree. The former kind of foodstuff (starch) is in them changed by peculiar diastatic ferments—the ptyalin of the saliva, and the amylopsin of the pancreatic juice—into sugar, which is taken up from the digestive tract, as such, by the radicles of the portal vein to be carried to the liver, there to undergo, by a process of dehydration, its conversion into glycogen. The latter —proteid matter—is, in them, converted into peptones by the action of the pepsin (in the presence of hydrochloric acid) of the gastric juice, and the trypsin (in the presence of sodium carbonate) of the pancreatic juice, and afterwards split up in the liver into a non-nitrogenous portion (glycogen) and a nitrogenous portion (probably urea). We have thus a starch diet and an albuminous diet, producing glycogen in the liver of vertebrates. Where does the analogy exist in the case of the Mollusca? Their food is essentially, in the majority of cases, a vegetable one, and forming the chief chemical constituents of plants are carbohydrates and proteids. The carbohydrates are starch, inulin, dextrin, and sugars, including glucose, cane-sugar, and various others. Proteids are there as protoplasm, aleurone grains, crystalloids, gliadin, vegetable fibrin, and a native albumen which is soluble in water, and coagulable by heat, and, in many respects, identical with animal albumen. And we have previously spoken of the action, in the Mollusca, of the salivary secretion and of the amylolytic ferment of the secretion of the "Mitteldarmdrüse" in converting starch into sugar; and we have already spoken of the action of the trypsin of the secretion of the "Mitteldarmdrüse in the Mollusca, as converting colloid proteids into crystalloid peptones. And that, in the Mollusca, glycogen is immediately formed from the ingesta, has been proved by Barfurth, who found that after three weeks' fasting it had disappeared from the liver of *Helix*, but that it reappeared in from nine or ten hours after feeding, and by Hammersten in the decrease in the amount of glycogen showed (already noticed) in the liver of *H. pomatia*, which had hibernated in a warm room. There seems, then, every reason for us to believe, in the face of no evidence to the contrary, that the mother-substance of glycogen in the Mollusca is the same as that in the Vertebrata. What its destination in the Mollusca may be, we have very little direct evidence to guide us to a safe conclusion. In the Vertebrates, it is no doubt devoted to the production of heat and muscle-energy. Broken up, in them, by a blood-ferment again into sugar, as the exigencies of the system demand, it is taken by the radicles of the hepatic vein to undergo katabolic changes in the tissues. And that it is used up during muscular contraction by them does not admit of a doubt, for it has been experimentally proved that all the glycogen disappears during movement. What about the Mollusca? Barfurth has demonstrated that the quantity present in the muscles of *Helix* is inversely proportional to their activity, and should Wooldridge's theory that the Vertebrate blood-ferment is lecithin prove to be a fact, we have here two things that unite together to show, disregarding the facts we have already mentioned, that glycogeny in the Mollusca and in the Vertebrata are far from being dissimilar, since lecithin is present in the Molluscan blood. But, whatever its destination in the Mollusca may be, it is a point well worthy of mention in connection with this, that a large amount of reserve material must be stored away in their tissues in some kind, if not in the form of glycogen, for the exigencies of the system during hibernation, and also for prolonged muscular contraction, since Simroth has stated that a small *Helix* can move along when burdened with a weight nine times its own, and Sandford has proved that a *Helix aspersa*, weighing one-third of an ounce, can draw along a horizontal plane a weight weighing seventeen ounces (fifty-one times its own weight), and that another of the same species, one-quarter of an ounce in weight, can drag a weight of two and a quarter ounces after it when moving along a vertical plane (nine times its own weight).

Glycogen is rather a difficult substance to prepare pure, but the steps of the process are as follows. The organ suspected to contain it is taken from a recently killed animal, cut up into pieces, and plunged into boiling water in order to destroy any ferment that may be present, then boiled for some time and filtered. The filtrate is allowed to get cool, and dilute hydrochloric acid and potassio-mercuric iodide are alternately added to precipitate any proteids that may be present in solution, and this is continued until no precipitate any longer obtains. Then it is filtered—if any glycogen be present the filtrate is clear and opalescent—and the glycogen is precipitated from the filtrate by adding 70 to 80 per cent. of alcohol in excess. The precipitate obtained is then washed with 60 and 90 per cent. of alcohol, afterwards with ether, lastly with absolute alcohol, then dried over sulphuric acid and weighed. Claude Bernard (born 1813, died 1878), a great French physiologist, was the first to discover this substance in the liver-cells of the Vertebrates and the Mollusca; he published his researches in 1857. Besides being known under the name of glycogen, it is often spoken of as hepatine, benardine, and zoamyline or animal starch.

THE CIRCULATORY SYSTEM.—The fluid in the Mollusca which corresponds to the blood of the higher animals is known as blood-lymph or hæmolymph. In it white-corpuscles or lymph-corpuscles float. In *Helix* the hæmolymph contains a respiratory substance termed hæmocyanin, which contains copper united with a proteid, and turns bright blue, when oxidised; in *Planorbis corneus* it contains hæmoglobin. The heart is surrounded by a pericardium, and situated near the kidney or nephridium in the dorsal portion of the body-cavity. The pericardium communicates with the kidney by a ciliated tube—the *reno-pericardial canal* or *nephridial tube*. The heart is composed of an *auricle* and a *ventricle*, separated from one another by an *auriculo-ventricular valve*. The *aorta* arises from the base of the ventricle, and immediately divides into an *anterior* and a *posterior aorta*. The anterior aorta passes into the prostoma under the head of the spermatheca and the first coil of the intestine, and ends in relation with the undersurface of the buccal mass, giving branches on its way to the foot, crop, salivary glands, tentacles, reproductive organs and integument, and also special branches to the supra-pedal gland and rectum; the posterior aorta passes into the visceral hump, and supplies the intestine, reproductive organs, and hepato-pancreas. There are no special capillaries, and the blood passes from the arteries into blood-spaces,—the *transition vessels* of Nalepa,—from whence, after supplying nutriment to the tissues, it is returned to the heart by way of the various venous sinuses.

On cutting open the pedal sinus the openings of these transition vessels into this sinus can be observed. There are four venous sinuses, two *pedal sinuses*, a *visceral* or *marginal sinus*, and a *pulmonary sinus* or *circulus venosus pulmonis*. The pedal sinuses run along the foot, one on each side; the *marginal sinus* runs

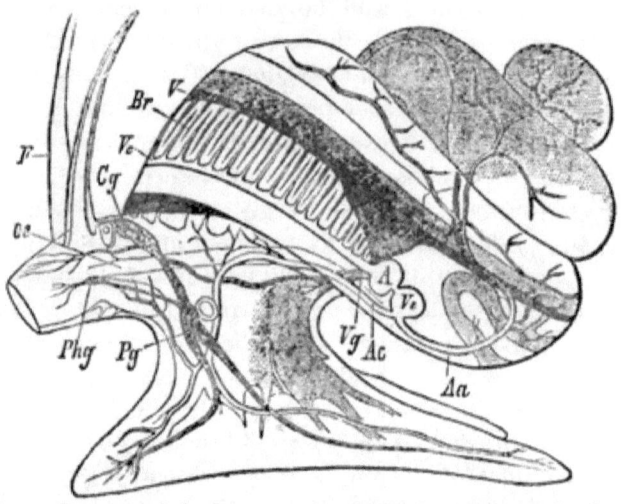

Fig. 9.—Nervous and circulatory systems of *Paludina vivipara* (after Leydig). *F*, tentacle; *Oe*, œsophagus; *Cg*, supra-œsophageal or cerebral ganglion, with eye; *Pg*, pedal ganglion with adjacent otocyst; *Vg*, visceral ganglion; *Phg*, infra-œsophageal or pharyngeal ganglion; *A*, auricle of heart; *Ve*, ventricle; *Aa*, posterior aorta; *Ae*, anterior aorta; *V*, veins; *Ve*, afferent vein; *Br*. gill.

from the apex of the spire to end in the *circulus venosus pulmonis*, a very large sinus which surrounds the base of the mantle cavity. From the last named sinus numerous afferent vessels—the *afferent pulmonary veins*—arise to form a plexus on the wall of the mantle cavity, and from these efferent vessels—the *efferent pulmonary veins*—carry the blood to the nephridium, where they break up into a plexus—the *renal plexus*—the venules of which join together to form a renal vein, which unites with a vein—the *pulmonary vein*—formed by the coalescence of a large number of efferent pulmonary veins. The pulmonary vein enters the auricle; it has been seen by Nalepa to pulsate rhythmically in *Zonites algirus*. The marine snail, *Haliotis*, has its heart somewhat similar to that organ in *Anodonta cygnea*. Its auricle is divided into two cavities, and its ventricle is pierced by the rectum. In *Fissurella* and in

Chiton this division of the auricle also obtains. In *Zonites algirus*, according to Nalepa, and in *Arion ater*, according to Jourdain, the transition vessels open into the venous sinuses by infundibular orifices. In *Zonites algirus* Nalepa has demonstrated a network of nerve-fibres in the auricle, ventricle, aorta, and larger vessels, but ganglia are rare. Haller has demonstrated ganglia in the cardiac walls of the marine snails, *Trochus, Haliotis*, and *Fissurella*. The arteries and arterioles in *Zonites algirus* are lined by an endothelium. "The venous system is represented in part by the cœlome (body-cavity), in part by vessel-like spaces, the walls of which are formed of homogeneous connective tissue with scattered nuclei, but which are not lined by epithelium." In *Chiton, Nerita, Neritina, Turbo, Emarginula, Fissurella*, and in *Parmaphorus* the ventricle is pierced by the intestine.* In *Janus* and *Emarginula*, besides the genera already mentioned, there are two auricles. In many Prosobranchs the heart is connected to the pericardium by fibrils. In *Aplysia* the commencement of the aorta is strengthened by muscle-fibres, a condition of things which, as Professor Owen says, resembles the bulbus arteriosus of fishes. In some forms, as *Pontolimax* and *Phyllirhoë*, there are no vessels except the aorta; in *Rhodope* and *Entoconcha* there are no specialised vessels at all. In *Patella* the "transition vessels" are separate from the cœlome. In *Haliotis, Fissurella*, and *Emarginula* the ventricle is double. Vulpian has found that nicotine, curare, and digitalis inhibit the heart's beat in about five or six minutes, just as they do in the higher Vertebrates. According to Ashford the pulse of *Helix hortensis* and *Hyalina cellaria* beats 12 to 14 per minute at 42 to 44° Fahr., 9 to 11 at 37 to 38° Fahr., and 4 to 8 at 26 to 33° Fahr.

THE RESPIRATORY SYSTEM.—In the Pulmogastropoda breathing is carried on through the pulmonary aperture which leads into the mantle-sac formed by an expansion of the body-wall (*mantle*). The walls of the mantle are very vascular, the vessels running in it being those already mentioned,—the afferent and efferent pul-

---

* Landsberg is opposed to the statement generally made that the intestine perforates the heart in *Neritina fluviatilis* and *Turbo rugosus*. In the former, according to his researches, the intestine passes between the heart and the nephridium; in the latter it pierces the pericardium only.

monary vessels, the pulmonary vein, and the circulus venosus pulmonis. External respiration can be readily performed then by this vascular network. In the branchiate Gastropoda respiration is performed by a *ctenidium* or gill, the water containing air being admitted in the majority of cases by an *anterior siphon*, and expelled by a *posterior siphon*. These gills can be well seen in *Paludina vivipara*. *Cyclostoma elegans* is an example of a branchiate becoming a pulmonate; it has lost its ctenidium, and respiration is now carried on by the walls of the pulmonary sac. It is the only English example of the PNEUMOCHLAMYDA, a class proposed by Professor Ray Lankester to contain all those terrestrial Mollusca in which the ctenidium has aborted, and respiration carried on by means of a lung. *Paludina vivipara* has a ctenidium, and consequently it belongs to an allied class termed the HOLOCHLAMYDA in which the ctenidium is present in the adult. The passage from the branchiate to the pulmonate form is well seen in the genus *Ampullaria*, where a ctenidium is to be found on the left side of the mantle-cavity, while the right side, which is separated from the left by a fold, is extremely vascular in character. The floor of the mantle is formed by a sheet of muscle tissue. Inspiration is effected by a contraction of this muscular layer, which bulges up when at rest into the mantle cavity just as our diaphragm does in our chest cavity; expiration is effected by the return of the muscular layer to a state of relaxation.

A somewhat similar condition of things present in *Ampullaria* has been found by Semper, it is interesting to note, in a terrestrial Crustacean, *Birgus latro*. According to Sabatier, the blood is driven, in *Ampullaria*, to the gills when the lung is collapsed, and to the lung when the gills are collapsed, by means of a valvular spur situated at the junction of the afferent vessel of the gills, and the afferent vessel of the lung. A few years back a good deal of discussion was taking place in the Continental magazines about the respiration of the *Limnæa* in deep water, and during winter when the ponds are covered with ice, the outcome of which is the interesting fact that under such circumstances they do not respire atmospheric air direct, but admit water to their pulmunary cavity instead.

THE NEPHRIDIUM.—The kidney or nephridium is situated in

the mantle-cavity between the pulmonary vein and the heart, and communicates with the pericardium by a ciliated canal,—the *reno-pericardial canal* or *nephridial tube*,—so that the pericardial cavity is in communication with the exterior through the ureter or duct of the kidney. The ureter runs alongside of the rectum in *Helix*, and opens near the anus into the mantle-cavity ; in *Arion* the kidney has "a simple round opening." The kidney is made up of two parts, one thin-walled and sacculated, the other yellowish, and possessing a lamellated structure. The excretion contains ammonium and calcium urate; and in *Hyalina* free uric acid and guanin in addition to these. If minutely examined, the orifice of the duct of the kidney in *Helix* will be seen to be carried as a groove—the *excretory groove*—over the anus, ending near the respiratory aperture.

THE NERVOUS SYSTEM.—Lying round the œsophagus is a cord of nervous tissue—the *circum-œsophageal mass*—which is divided into that portion which lies above the œsophagus (*supra-œsophageal ganglia*) and that portion which lies below the œsophagus (*sub-* or *infra-œsophageal ganglia*). The supra-œsophageal ganglia are two in number, and are united together in the middle line by a band of nerve-fibres (*commissure*); from them nerves pass to nnervate the lips, tentacles, and the buccal mass. The infra-œsophageal ganglia are divided into two portions—an anterior portion (*pedal ganglia*) and a posterior portion (*viscero-pleural ganglia*)—by the passage of the anterior aorta through the centre of its mass. The pedal ganglia give branches to the muscular substance of the foot; the viscero-pleural ganglia send branches to the body-wall and viscera. A miscroscopical examination of a ganglion shows it to consist of nerve-cells generally unipolar, sometimes bipolar and multipolar, placed round a central mass of nerve-fibres derived from them, and arranged in a reticulated manner. To this central mass Leydig has given the name of *Punkt-substanz*.

The *nerve to the tentacle* arises from the supra-œsophageal ganglion with the labial nerves, and ends at the summit of the tentacle by a knob which consists of nerve-cells and nerve-fibres, surrounded by a muscular sheath derived from the retractor muscle of the tentacle, and which is thought to be olfactory in function, since on excising it on both sides the snail no longer recoils from strong smelling liquids such as turpentine, which it invariably does when one or both are present. In *Limnæa stagnalis* and *Planorbis* the infra-œsophageal ganglia are orange or red-coloured, and very pretty indeed. The nervous system of *Limnæa stagnalis* well dissected out is the prettiest dissection I

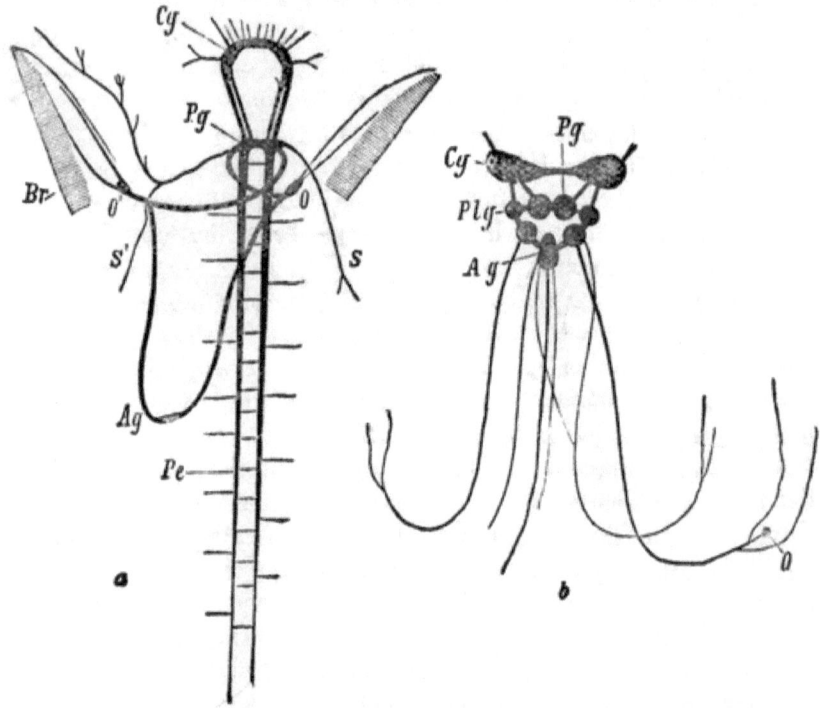

Fig. 10.—*a*, Nervous system of Haliotis (diagrammatic after Sprengel); *Cg*, circum-œsophageal mass; *Pg*, pedal ganglion; *Plg*, pleural ganglion (commissural ganglion); *Ag*, abdominal ganglia; *O* and *O'* olfactory organs; *Pe*, pedal cord; *S* and *S'* lateral nerves; *Br*, gills; *b*, nervous system of *Limnæa*, (after Lacaze-Duthiers).

have ever seen. The circum-œsophageal mass in the large *Tritons* is almost cartilaginous to the touch.

THE OLFACTORY ORGANS.—The sense of smell may exist in the supra-pedal gland, the dorsal tentacles, or in the lobate processes around the mouth. An *osphradium* is to be found in *Limnæa*, *Planorbis*, and *Physa* consisting of ganglion cells in relation with a depression in the mantle-chamber above and behind the respiratory aperture. A nerve supplies it in *Limnæa* from the right visceral ganglion: in *Planorbis* and *Physa* the nerve comes from the left visceral ganglion. In the *Helices* the osphradium is apparently absent, but a nerve has been described by Sarasin in *Helix personata* which arises from the right visceral ganglion, and ends near the pulmonary orifice in a collection of ganglion cells. The same nerve has been found in *Limax cinereo-niger, Succinea amphibia, Bulimus detritus* and *Bulimus decollatus*, but the ganglion cells are absent.

THE AUDITORY ORGANS.—Before ever the auditory organs were dissected out the late Dr. Grant surmised their existence, for he could not but think that the sounds emitted by *Tritonia arborescens* under water, were intended to be heard by its fellows. Siebold first discovered the auditory organs or, as they are generally termed, the *otocysts*. They are a pair of minute white bodies in very close relation with the pedal ganglia, and in *Paludina* are movable by muscles. Each consists of a connective tissue capsule containing fluid, and a large number of calcareous granules (*otoliths*) which are in constant motion. These otoliths, disturbed by the sound waves from a sounding body, strike upon the nerve filaments of the auditory nerve, which thus communicates a wave of change to the supra-œsophageal ganglion, the result of which is the perception of sound by the animal.

THE ORGANS OF SIGHT.—We have previously considered the relation which the eyes may bear in different species to the dorsal tentacles. The eye consists of a cornea, lens, or vitreous body (Carrière), and a retina composed of pigmented and non-pigmented cells. The lens or vitreous body is structureless, and fills the whole of the cavity of the eye; the cornea is composed of connective tissue, with a layer of transparent cells on its outer surface The optic nerve is derived from the nerve to the tentacle in the STYLOMMATOPHORA; in the BASOMMATOPHORA it is an independent nerve. Snails cannot accommodate for long distances, since they cannot distinguish objects till within a quarter of an inch from the eyes. In *Chiton* and *Vermetus* the eyes are absent. In *Patella* the eye is cup-shaped. The eye can be invaginated into the body by the retractor muscle of the tentacle which is inserted into the columella.

ANATOMY AND PHYSIOLOGY OF A SNAIL.

THE REPRODUCTIVE ORGANS.—The reproductive organs of the typical Pulmonate Gastropod may be well divided into three groups, a male group, a female group, and a group common to both the male and female group. The male group contains the *penis sac, vas deferens, prostate,* and *flagellum;* the female group is made up of the *oviduct, dart-sac, receptaculum seminis, albuminiparous gland,* and *finger-shaped glands;* the group common to both the male and female portions contains the *hermaphrodite gland* or *ovotestis,* the *hermaphrodite duct,* the *common generative canal,* and the *cloaca* or *vestibule.* These various parts will be readily seen at a glance in the accompanying figure of the generative organs of our largest English land-snail, *Helix pomatia.* The principal organ is the *hermaphrodite* gland or *ovotestis* situated in the left lobe of the "Mitteldarmdrüse," and consisting of numerous follicles connected together by connective tissue. The ova and spermatozoa are formed from the wall of the follicle and from identical cells, but they are developed at different times,—the former from the cells on the outer wall, the latter from its more central portion. A

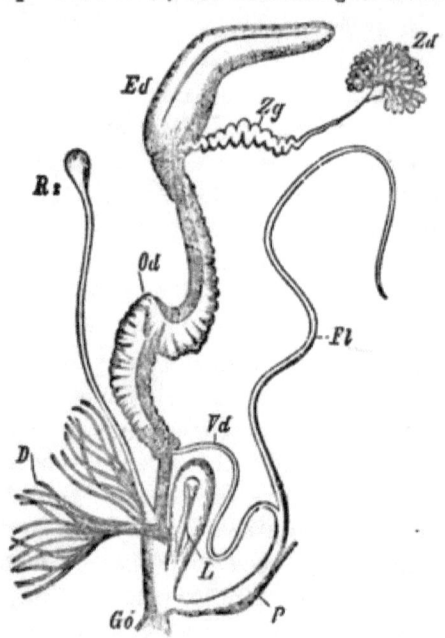

Fig. 11—Reproductive organs of *H. pomatia.* Zd, hermaphrodite gland; Zg, its duct; Ed, albuminiparous gland; Od, common generative canal; Vd, vas deferens; P, protrusible penis; Fl, flagellum; Rs, receptaculum seminis; D, finger-shaped gland; L spiculum amoris in the dart-sac; Go, common generative opening (after Baasen).

spermatozoon is a filiform body—the filamentous portion being known as the *tail*—with an enlargement at its anterior extremity (*head*); an *ovum* is a rounded cell without a distinct outer coat, as in the higher animals, and it consists of protoplasm cotaining a more or less rounded structure (*germinal vesicle*), which in its turn encloses a larger, more refractile body (*germinal spot*). It is a primitive undifferentiated cell pure and simple; the germinal vesicle represents the nucleus, the germinal spot the nucleolus. We shall have to refer to this ovum more in particular directly, when we come to consider the chief steps in the development of the snail.

The ducts of the follicles of the ovotestis unite to form a duct common to them all, lined with non-ciliated squamous or columnar epithelium—the *hermaphrodite duct*—which leaves the gland, courses through the albuminiparous gland, and at last, after receiving the duct of that gland, leaves it to form the *common generative canal*. The common generative canal is composed of the *vas deferens* and *oviduct* united together. The lumen of the vas deferens is in communication with that of the oviduct by numerous slits in the dividing membrane, and around the vas deferens a somewhat fluffy-looking mass will be seen on a minute examination—the so-called *prostate*. This common generative canal, however, at last divides into a longer independent portion —the *vas deferens proper*—which belongs to the male organs of generation, and into a shorter independent portion—the *oviduct* —which constitutes part of the female generative organs. The common generative canal is lined with a ciliated epithelium, and its walls are glandular, the independent portion of the oviduct is ciliated likewise, but in the vas deferens proper the cilia are absent. The *albumen* or *albuminiparous* gland is a greyish-white or yellowish tongue-shaped organ, situated between the stomach and the right lobe of the "Mitteldarmdrüse." It is a racemose gland, the ducts of the acini of which open into a common duct —the *duct of the albuminiparous gland*—which has its epithelium ciliated, and which opens into the hermaphrodite duct just before it leaves the gland as the common generative canal. There may be two accessory albumen glands, as in *Limnæa stagnalis*, or one, as in *Limnæa peregra*.

The vas deferens proper runs between the female generative organs and the buccal mass, lies on the back of the crop, and terminates in the *penis sac*. It gives off a long coiled diverticulum—the *flagellum*—previous to its termination. Sometimes this flagellum is given off from the base of the penis. The penis sac during copulation is exserted; attached to it is a *retractor muscle*. The flagellum, with the posterior part of the penis, secretes the *spermatophore* or *capreolus*. This is a body formed of mucus with its edges folded inwards, so as to form a groove in which spermatozoa becomes lodged, and more or less, as it were, glued together into a mass. When the flagellum is absent the spermatophore may be formed in the vas deferens (*Hyalina*) or in the penis (*Arion*). The flagellum is absent in the American Helices. In *Loligo vulgaris*, and in many other forms, a special pouch is developed in the vas deferens for the spermatozoa; this is known as *Needham's pouch*. According to the recent researches of Ashford the flagellum of *Testacella haliotidea* has a special *retractor muscle*.

The oviduct, like the flagellum, gives off in its course a diverticulum—the *spermatheca* or *receptaculum seminis*—the function of which is to receive during coition a spermatophore derived from another snail. This structure consists of a globular head—the *Swammerdamian vesicle*—lying near the albumen gland, and of a duct—the *duct of the spermatheca*. In *Helix aspersa*, and in other members of the genus to which it belongs, there is an additional diverticulum given off which ends near the aorta, and may possess a globular head. Below the point of origin of the spermatheca the oviduct is generally spoken of as the *vagina*; this runs forwards to open side by side with the penis into a *common genital cloaca* or *vestibule*. In the genera, *Limnæa*, *Planorbis*, and *Ancylus* the male aperture (*penial aperture*) is distinct from the female aperture (*os vaginæ*), the former being the more anterior of the two. The consequence is that, anatomically, at any rate, self-fecundation is possible in them, and this has been actually observed in *Limnæa auricularia*. Into the vagina open two structures,—the *mucous* or *finger-shaped gland*, and the *dart-sac*. The former of these consists of a pair of tufted tubular glands, which secrete a highly refractile fluid that is poured out when coition takes place. The dart-sac is a pear-shaped organ, containing in its interior a peculiar spicule, known as the *dart* or *spiculum amoris*, which is of a calcareous nature, except in *Vitrina elongata*, where it is chitinous.

The dart varies in shape in different species. It is to be found only in the *Helices** and in the slugs *Agriolimax maculatus* and *Tebennophorus*. Mr. Ashford has made the admission that the dart, in his thinking, is not thrust out of the body; but he is wrong, for it is forcibly ejected from one snail to the other before they engage in *coitus*, and it is not a rare thing, when doing field work, to pick up one of them from out of the grass. He based his statement on the fact that when dissecting a dart out of its sac it is always to be found connected at its base to the posterior end of the lumen of that organ; the mistake is here, since the dart is secreted continuously by cells in the posterior end of the sac, and this causes it to simulate an appearance as if attached. The function which this dart serves in the economy of the snail has not, as yet, been thoroughly investigated. My friend, Professor J. Bland Sutton, suggested to me that it is an organ of cuticular irritation, and I find that Simroth holds the same view; but I cannot see any good reason to make this hypothesis at all stable, for, if so, why is it not a constant feature in the anatomy of the majority of snails? To me, the most probable and the most feasible view, as yet, has been advanced by another friend of mine, Mr. W. E. Collinge of Leeds. He believes that in it we see the vestige of a structure which was once constant in all snails, and which was used in former times as a weapon of defence. Mr. Collinge is still collecting together facts to support his view, and we shall all await the result with interest. In *H. hispida*, *H. rufescens*, *H. plebeia*, and *H. villosa* there are two dart-sacs, each containing a spiculum amoris. The Gastropoda, with the exception, perhaps, of *Limax lævis*, are, as is seen from the foregoing description, monœcious (Gr., *monos*, single; *oikos*, house) or hermaphrodite (Gr., *Hermes*, Mercury; *Aphrodite*, Venus) animals. The generative act consists essentially in an exchange of spermatophores, which are said, in *H. pomatia*, to resolve in about ten days after the receival by the animal. The spermatozoa set free by this resolution fertilise the ova in the oviduct as they are descending from the ovotestis. By the land-snails the eggs are laid singly in the earth or under stones and logs of wood; by the pond-snails they are laid in an

* It is rare in the American Helices; so are the finger-shaped glands.

ANATOMY AND PHYSIOLOGY OF A SNAIL.

elongated or ovoidal mass (*nidamental mass*), which is fastened to a stem or a leaf of an aquatic plant, or often to stones and bricks that lie on the bed of the pond or river. The *Bulini* make a nest for their eggs by cementing together leaves of trees. *Paludina vivipara* and *P. contecta* are ovoviviparous among water-snails; *Helix rupestris*, a few species of *Clausilia* and *Pupa*, a *Vitrina* and an *Achatinella* are viviparous among land-forms.

The different processes which the egg undergoes in its development cannot be described here in detail, because they are of too complicated a character to be well understood by the beginner. Enough, however, will be stated to render the reader familiar with the more striking changes, and should he wish to advance

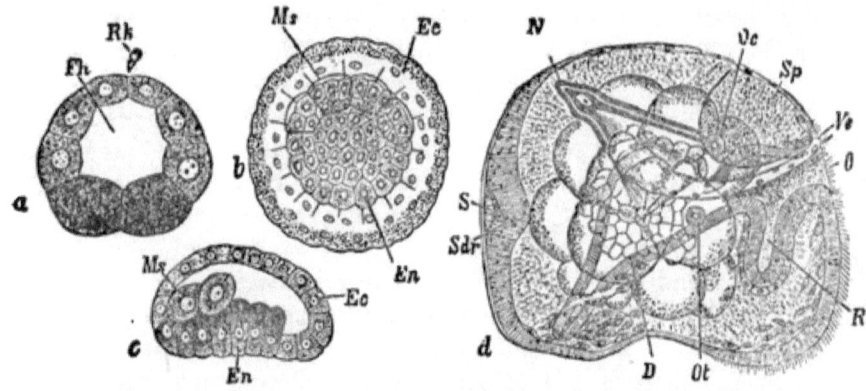

Fig. 12.—Some stages in the embryonic development of *Planorbis* (after C. Rabl). *a*, optical section through a segmenting ovum (24 segments). *Rk*, polar bodies; *Fh*, segmentation cavity, *b*, stage with four mesoderm cells viewed from the vegetative (lower) pole; *Ms*, mesoderm cells; *En*, endoderm; *Ec*, ectoderm, *c*, oblique optical longitudinal section through the stage with four mesoderm cells, d, older embryo in which the shell-gland has shifted to the right, *Sdr*, shell-gland; *s*, shell; *O*, mouth; *D*, alimentary canal; *R*, commencing radula; *Sp*, apical plate (thickening of præoral lobe); *Oc*, eyes; *ot*, otolith; *N*, primitive kidney; *Ve*, velum.

further into the subject he may consult the embryological text books in general use, such as Balfour's "Comparative Embryology," and Haddon's "Introduction to the Study of Embryology," or the article "Mollusca" by Professor Ray Lankester in the sixteenth volume of the last (ninth) edition of the *Encyclopædia Britannica*. Taking *Limnæa stagnalis* as a typical instance, the ovum first divides into two and then into four portions, each division being preceded by the division of the germinal vesicle. Then from the upper surface of these four spheres which are known as *macromeres*, numerous smaller spheres (*micromeres*) are budded off. These go on increasing in number to a large extent, macromeres, and micromeres being produced by the division of those previously existing. Meanwhile a cavity is formed in the centre of their mass,—the *segmentation cavity*,—and the macromeres, becoming invaginated or tucked into the segmentation cavity, form a double-layered embryo, the external wall (*ectoderm*) being formed of micromeres, the internal wall (*endoderm*) being made up of macromeres. This stage is known as the *gastrula-stage*, and the opening produced by the invagination of the macromeres, the *blastopore*. This blastopore closes as the larva advances in its development, and the place where the future mouth and anus are to be formed becomes marked off by invaginations of the ectoderm occurring at the anterior and posterior ends of the closing slit as the *stomodæum* and *proctodæum* respectively. The embryo now becomes oval in contour and surrounded by a circlet of cilia,—the "*Trochosphere-stage*" of Lankester,—which divides the body into two unequal hemispheres, a smaller, which develops into the visceral mass, and a larger, which becomes the foot. The ciliated circlet then becomes the *velum* —a structure identical with the disc of a Rotifer—the foot grows larger, and becomes bilobed, the mantle and shell become developed, the tentacles begin to appear, and the snail passes from this "*veliger stage*" into a definite molluscan condition, the whole process of development from the egg to the mature state occupying a period of time varying from twenty to thirty days.

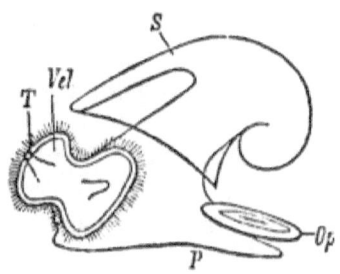

Fig. 13.—Older larva of a Gastropod (after Gegenbaur). *S*, shell; *P*, foot; *Vel*, velum; *T*, tentacles; *Op*, operculum for the closure of the shell opening.

CHAPTER III.

THE ANATOMY AND PHYSIOLOGY OF A FRESH-WATER MUSSEL.

TAKE a shell of the Swan Mussel (*Anodonta cygnea*).

(A) Examine its outer surface. Note :—

1. Its oval form, blunt at its anterior, but produced at its posterior extremity.
2. The *umbo*, a small blunt prominence near the dorsal border.
3. The *lines of growth*, running concentric to the umbo.
4. The *ligament*, uniting the two valves together along their dorsal margins. "The relations of the ligament to the shell valves show that strictly speaking the valves ought to be regarded as parts of a continuous structure. The dorsal region of this structure does not undergo calcification, or only to a very slight extent, inasmuch as the economy of the animal requires that it should remain flexible. It is an adaptation of an originally univalve shell."

(B) Examine its inner surface. Note :—

1. The *anterior adductor impression*, near the anterior end of the shell.
2. The *anterior retractor impression*, a small area continuous with the anterior adductor impression.
3. The *protractor impression*, about $\frac{1}{8}$th of an inch posterior to the inferior portion of the anterior adductor impression.
4. The *posterior adductor impression*, an oval depressed area near the posterior end of the shell.
5. The *posterior retractor impression*, a small depression continuous with the posterior adductor impression.
6. The *pallial line*, an even curved line running parallel to the inferior margin of the shell. In *Anodonta* this line is even throughout its whole extent, and the shell is said to be *integri-*

palliate; those shells in which the pallial line is not evenly described, but forms a sinus in relation with the posterior adductor impression are said to be *sinupalliate.*

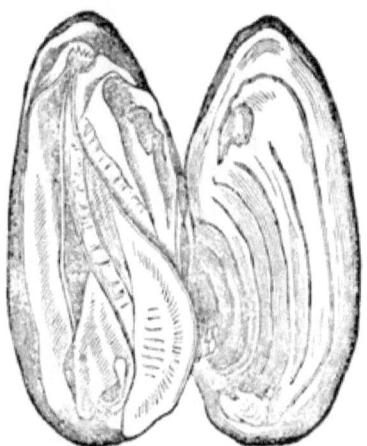

Fig. 14.—The anatomy of a Mussel, and the internal markings of its shell.

7. The *lines of shifting of the muscles,* numerous lines running from the different impressions towards the umbo.

8. The *hinge,* a longitudinal ridge running along the dorsal

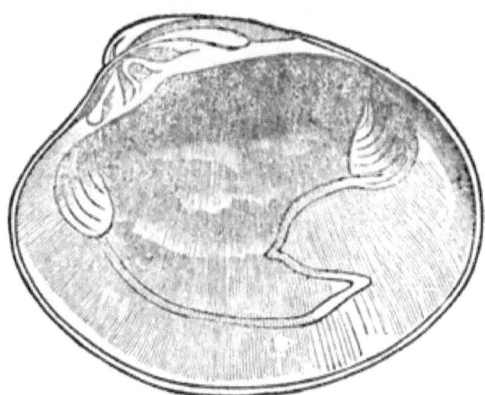

Fig. 15.—Right valve of the shell of *Cytherea* showing internal markings.

portion of each valve. The absence of teeth in the Anodon shell; in *Unio, Sphærium,* and *Pisidium* teeth obtain. The teeth consist of three sets: *cardinal teeth* situated below the umbones, *anterior lateral teeth* in front of them, and *posterior lateral teeth* behind them. *Dreissena polymorpha* may, or may not, possess teeth.

9. The *ligament* connecting the two valves of the shell together. In *Anodonta*, according to F. Müller, this ligament consists of an outer part and an inner part, the former being continuous with the epicuticula of the shell, the latter having a distinct ridge—the homologue of that portion produced into teeth in *Unio*, *Sphærium*, and *Pisidium*, where it borders the nacreous layer. This ridge, like the teeth, gives attachment to muscles from the foot. The

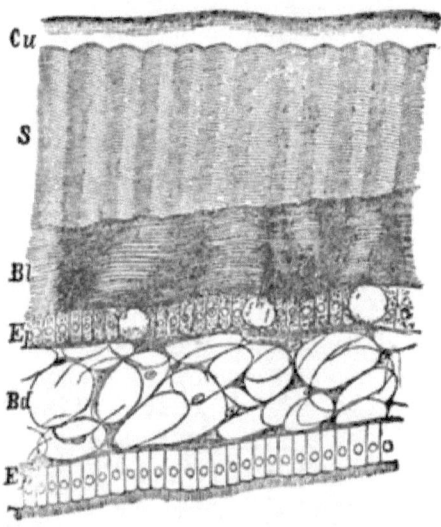

Fig. 16.—Vertical Section through the shell and mantle of *Anodonta* (after Leydig). *Cu*, cuticle, *S*, prismatic layer; *Bl*, nacreous layer; *Ep'*, external epithelium of mantle; *Bd* connective tissue substance; *Ep''*, internal epithelium of mantle.

outer part is laminated; the inner part is radially striated, and consists of radial fibres made up of two different kinds of substances which refract light differently.

In microscopic structure the shell may be said to consist of three layers,—an *epicuticula* or *periostracum*,* a *prismatic layer* and a *nacreous layer*. The periostracum is thin, horny, laminated, and consists of conchiolin; the prismatic layer consists of numerous polygonal prisms united together by conchiolin and

* Often erroneously called the epidermis.

placed obliquely to the surface of the shell; the nacreous layer consists of calcareous laminæ alternating with organic laminæ. It is formed by the cells of the general mantle surface, the epicuticula by cells near the mantle-edge, and the prismatic layer by cells posterior to those forming the epicuticula. Müller believes the shell to be alive, growing by intussusception, or the deposition of particle between particle; while Tullberg seems to hold the view that the organic matter is produced by fibrillation of the cells of the mantle. The calcareous portion of the shell consists in the main of calcium carbonate, with, sometimes, traces of calcium phosphate, alumina and silica. About one-third of the animals of *Anodonta cygnea* which I have examined were the subjects of what I may term the *pearly di...hesis*. Pearls are formed by depositions of nacre round a grain of sand which has gained admission into the pallial cavity, provided, I would say, that the organism of the animal be predisposed, since sand-grains, by the very nature of their habitat, must gain admission into the mantle cavity of all mussels indiscriminately, while it is only in about a third that the grains of sand become surrounded with nacre. It is a point well worthy of mention that when pearls occur they are generally situated in and around the pericardium.

The shell of *Sphærium corneum* has no prismatic layer, and the nacreous layer has a reticulated structure. The shell is caniliculated, and into each canal a process of the mantle extends. The structure of the shells of our other Lamellibranchs are worthy of investigation, and I would recommend it as a study to those of my readers more especially interested in the anatomy of the Mollusca.

Examine the animal within its shell. Note :—

1. The *mantle* or *pallium* divided into two lobes, a right and a left. Its thickened ventral border, the *pallial muscle*.

2. The *gills* or *ctenidia*, lying internally to the mantle, two on each side of the body.

3. The *labial palps*, two pairs of triangular folds in front of the gills.

4. The *foot and visceral mass*, a large ploughshare-shaped mass lying between the right and the left mantle-lobes.

5. The *anterior* and *posterior adductor muscles* passing from

one valve of the shell to the other. In *Anodonta* they are equal in size (Isomya); in *Dreissena polymorpha* the anterior is smaller than the posterior adductor muscle (Heteromya). Some Lamellibranchs have only one adductor muscle, and are then called "*monomyaries;*" those with two adductor muscles are termed "*dimyaries.*"

6. The *anterior* and *posterior retractor pedis muscles*, near the anterior and posterior adductors.

7. The *protractor pedis muscle*, behind the anterior adductor.

8. The *pallial muscle*, attaching the mantle to the pallial line.

9. The *inhalent aperture*, bounded with tentacular fringes at the hinder end of the body.

10. The *exhalent* or *cloacal aperture*, immediately dorsal to the inhalent aperture.

Thus Keferstein's analogy of the general structure of a Lamellibranch to a book is appropriate. The cover of the book is the valves, the back is the hinge-line, the fly-leaves are the mantle-lobes, the second and third pages on each side are the gills, and the interior of the book is represented by the visceral mass and foot.

THE MANTLE, MANTLE-CAVITY, AND GILLS.—The mantle-lobes are symmetrical. They are continuous with the body-wall above, below they are free. The epithelium on their internal surfaces is ciliated. Between them lies the *mantle* or *pallial cavity*, divided by the base of the gills into an upper or *suprabranchial chamber*, and a lower or *infra-branchial chamber*. In this pallial cavity lie the gills, foot, labial palps, and the greater part of the visceral mass. The gills or ctenidia consist of a pair of plates lying on each side of the foot. Each plate is made up of a descending and an ascending lamella, the descent and ascent being from a ridge on the body-wall termed the *gill-axis*. Each lamella is composed of a number of gill-filaments which have become fused together so as to form a trellis-work-like structure. It will be best to trace these gill lamellæ from the gill-axis. From the gill-axis two descending lamellæ proceed, one to form the inner lamella of the outer gill, the other to form the outer lamella of the inner gill. These, descending for some distance, at last turn sharply upon themselves, and the inner lamella of the outer gill thus becomes the outer lamella, and the outer lamella of the inner gill the inner lamella,—*i.e.*, of course, of their respective gill-plates. The outer lamella of the outer gill-plate becomes united by concrescence with the mantle, the inner lamella of the inner gill remains free as a slit through which the infra-communicates with the supra-branchial chamber. Thus, "regarding, then"—as Professor Lloyd Morgan in that most practical of all recent text-books, "Animal Biology," says,—"the gills as composed of a greater number of filaments with descending and ascending moieties, and we are justified by comparative morphology and development in so regarding them, we have to note the large amount of fusion or concrescence of parts.

1. The inner surfaces of adjoining filaments have become united by a membranous expansion, the lamellar membrane,

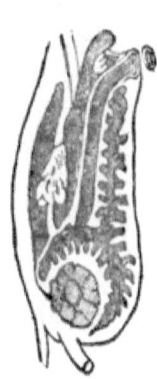
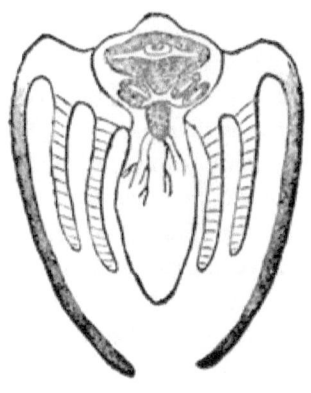

Fig. 17.—The Organ of Bojanus in diagrammatic section.

Fig. 18.- Diagrammatic Section of a Mussel.

perforated with oval fenestræ or windows. Thus the ascending and descending filaments respectively come to form ascending and descending lamellæ.

2. The two lamellæ of each gill become fused by interlamellar junctions, occurring in the inner gill at intervals of about twenty filaments, and in the outer gill at intervals of seven or eight filaments.

3. The ascending lamella of the outer gill is concrescent with the mantle along its whole length

4. The ascending lamella of the inner gill is concrescent anteriorily with the foot.

5. The ascending lamellæ of the inner gills are concrescent with each other posteriorily.

6. The union of the gill-axis with the body-wall may be partly due to concrescence."

The surfaces of the gill-filaments are covered with cilia, which may be divided into three groups, viz.,—those with the shortest cilia (*lateral*), those with medium-sized cilia (*frontal*), and those consisting of a single row of cells which have very long cilia (*lateral frontal*). The lashing of these cilia produces a current of water, which passes in the supra-branchial chamber by the inhalent aperture, and, after finding its way into the infra-branchial chamber through the fenestræ in the gill-lamellæ and the slit between the visceral mass and the inner gill, leaves the body through the exhalent aperture. Professor Ray Lankester thinks that perhaps the gills and labial palps are the homologues of the præ-and post-oral ciliated bands of the Echinoid and Ophiurid larva, *Pluteus*, or the *Tornaria* larva of *Balanoglossus*. Doubtless, as Peck thinks, the current of water produced by these cilia subserves the function of alimentation as well as that of respiration.

THE DIGESTIVE SYSTEM.—The mouth is situated below the anterior adductor muscle, and is bounded by a pair of membranous folds on each side—the *labial palps*—which lessen in size the nearer they are to the mouth. The labial palps at the mouth are confluent on either side, and their free edges encircle a groove, known as the *ciliated groove*, which runs into the mouth-aperture. We have just referred to what Professor Lankester has suggested regarding these palps taken together with the ctenidia; it would be as well to mention that Lovén has suggested that the palps are the remains of the velum. The œsophagus is short but spacious, and leads back into a dilated chamber of an irregular shape—the *stomach*—which lies in the substance of the digestive-gland in a position in the body immediately anterior to the pericardium. The intestine commences on the floor of the stomach by a well-marked pylorus, runs downwards and backwards towards the posterior end of the body in intimate relation with the generative gland, then doubling on itself, it passes again downwards and backwards to the mid-ventral region, where it redoubles upon itself, and after running for a little distance pierces the anterior end of the pericardium, from whence it runs as a straight rectum through the ventricle of the heart, ending by an anus on the dorsal side of that posterior portion of the supra-branchial chamber which is sometimes spoken of as the cloacal chamber. The rectum has a well-marked typhlosole running its whole length, the function of which is probably to increase the absorptive area. In the walls of the intestine of most Lamellibranchs, or in an appended cæcum, a peculiar crystalline style is often found, the formation and function of which have been made the recent subjects of research by Hazay. It is an albuminoid body, which probably acts as a stopper to the pylorus. Its formation takes place from spring to autumn, and is generally completed by October by the differentiation of a gelatinous mass in the stomach, which is apparently food-material. The style is, however, absorbed during the winter months, and the question naturally occurs to us whether or not we must look upon it as reserve nutrient material. The cells lining the intestinal tract are ciliated. The liver, digestive gland, or hepato-pancreas lies between the anterior adductor muscle and the pericardium, and on either side of the stomach, into which it pours its secretion through several ducts. It secretes trypsin and a diastatic ferment, and the extract of the gland is active in either an acid, a neutral or an alkaline medium. There are no calcareous cells, the cells present being granular and ferment cells.

THE ORGANS OF CIRCULATION.—The heart consists of two *auricles* and a pear-shaped median *ventricle*, situated in a *pericardium*, which communicates with the glandular portion of the nephridium by several *reno-pericardial apertures.* The pericardial cavity is pierced by the intestine, the ventricle by the rectum. Between each auricle and the ventricle there is an *auriculo-ventricular valve* consisting of pocket-shaped segments, which only admit the blood from the auricle into the ventricle. From the anterior end of the ventricle the *anterior aorta* arises and runs along the dorsal surface of the rectum ; from its posterior end the *posterior aorta* arises and runs along the ventral surface of the rectum. The anterior aorta passes into the visceral mass behind.

ANATOMY OF A FRESH-WATER MUSSEL. 53

the anterior retractor muscle, and divides into a *visceral* and a *pedal artery*, the former of which supplies blood to the digestive and generative organs, the latter to the foot, besides giving off a *labial branch* to the labial palps, and a *pallial branch* to the mantle and anterior adductor muscle. The posterior aorta gives branches to the mantle, body-wall, and posterior adductor muscle. These arteries break up into numerous lacunar spaces, from whence the blood finds its way into the *vena cava* which lies under the pericardium. From this vessel the blood passes to the organs of Bojanus, in which the blood-vessels form an irregular plexus, and thence by the afferent branchial vessels to the gills, where it becomes oxidised, losing carbon dioxide and gaining oxygen. From them, the now arterialised blood is carried by the efferent branchial vessels to the efferent branchial sinus, which expands to form the auricle. The direction which the blood takes may then be arranged in a tabular form, thus:—

Fig. 19.—Diagram of the heart of a Mussel.

1. Ventricle.
2. Anterior and posterior aortæ.
3. Arterioles and lacunar spaces.
4. Vena cava.
5. Organ of Bojanus.
6. Gills.
7. Auricles.
8. Ventricle.

The blood or hæmo-lymph consists of a colourless plasma, in which colourless corpuscles (*leucocytes*) float. The extension of the foot for the purposes of locomotion is due to a congestion of the lacunar spaces in that organ. The vein which conveys the blood from the foot is surrounded by a muscular sphincter, which, when the animal wishes to walk, contracts, and so prevents any return of blood, while the heart continues at the same time to pump blood into the lacunar spaces. A peculiar plexus of vessels, known as the *pericardial gland* or *Organ of Keber*, exists near the anterior end of the pericardial cavity. Its function is unknown.

THE EXCRETORY ORGANS.—The *excretory organs, organs of Bojanus*, or *nephridia*, are situated just beneath the pericardium. They excrete guanin, not uric acid. Each is composed of a glandular part, communicating in front with the pericardial cavity, and of a non-glandular muscular part or *vestibule*, which communicates with the fellow of the opposite side by the *inter-renal aperture*, and with the exterior by a small pore—the *external renal aperture*

Fig. 20.—The Organ of Bojanus in diagrammatic plan.

—which opens into the supra-branchial chamber dorsal to the generative opening. The glandular portion consists of numerous lamellæ, and on these Kollmann has recently described blind ciliated funnel-shaped openings which may in *Anodonta* be present to the number of two hundred.

THE NERVOUS SYSTEM.—The nervous system consists of three pairs of ganglia—the *cerebral* or *supra-œsophageal* (which according to Sprengel represents the cerebral plus the pleural ganglia in the snail), the *pedal*, and the *osphradial, olfactory*, or *parieto-splanchnic ganglia*, with numerous commissures. The *cerebral ganglia* are situated at the base of the labial palps, and are united together over the mouth by an *inter-cerebral commissure;* they are triangular in shape, and supply branches to the palps, mantle, gills, and neighbouring muscles. From these ganglia two cords, forming the *cerebro-pedal commissures*, run downwards and backwards to the *pedal ganglia* in the foot. The pedal ganglia are situated at the junction of the visceral mass and the foot; they are orange-coloured bodies giving branches to the foot. It is generally stated that the nerve to the otocyst comes from the pedal ganglia, but according to the recent researches of Simroth, it must be considered as arising from the cerebro-pedal commissure.

From the cerebral ganglia, two cords also proceed as the *cerebro-splanchnic commissures* through the substance of the nephridium

Fig. 21.—Diagram of the nervous System of a Mussel.

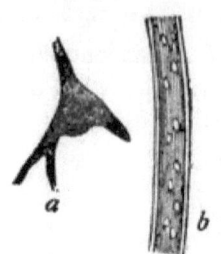

Fig. 22.—*a*, Ganglionic or nerve cell with nucleus; *b*, nerve-fibre showing contained granules.

to two yellow *parieto-splanchnic ganglia* on the underside of the posterior adductor muscle. These ganglia send branches to that muscle, and to the foot.

THE ORGANS OF SPECIAL SENSE.—Many years ago Siebold discovered in the foot of *Sphærium corneum* a "small sacculus," which contained "a cretaceous nucleus of a crystalline structure performing remarkable oscillatory movements," and this he very properly regarded "as a rudimental organ of hearing." In *Anodonta* this auditory organ or *otocyst* is situated near the pedal ganglia, and it has its nerve from the cerebro-pedal commissure. Lankester thinks that an organ described by him as the *osphradium*, consisting of a layer of elongated epithelial cells in relation with the osphradial ganglion, subservient to the sense of smell, while the osphradial ganglion itself is considered by some as a co-ordinating centre for sensations received from the inhalent current. The labial palps may be organs of taste; the general surface of the animal is sensitive to touch.

THE REPRODUCTIVE ORGANS.—With the exception of a few, such as *Pisidium* and *Sphærium*, all the Lamellibranchs are diœcious, a condition of things which was first made known by Leuwenhock. *Anodonta*, however, is sometimes found to be monœcious. The generative gland is a racemose gland situated in the visceral mass. The ovary cannot well be distinguished from the testis, except upon a microscopical examination, although the former portion is more reddish than the latter. The genital aperture opens just below the orifice of the ureter. The ova in the female do not pass directly out of the body, but congregate in large numbers between the two lamellæ of the outer gill, where they develop into peculiar larval forms—which were considered

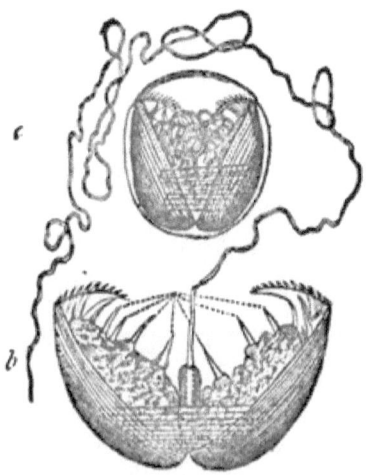

Fig. 23.—A, Glochidium-larva still within the egg; B, Glochidium viewed from the side showing the teeth on its valves.

parasites of the mussel by Rathké and Jacobson—known as *Glochidia*. Each valve of a glochidium-shell is shaped like an

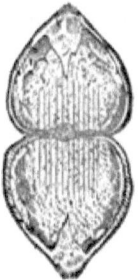

Fig. 24.—Shell of Glochidium widely opened, showing the single adductor muscle.

equilateral triangle, with the apex incurved, and produced into a sharp saw-edged tooth. A single adductor muscle is present;

the foot is small and slightly developed, and in its place two filaments—the *byssal filaments*—are seen projecting from the larva.

In this form the embryo is ejected from the gill of its broodmother into the water. Then, sinking down to the bottom, the shell gapes widely, for the single adductor muscle is not strong enough to keep the valves together. Swimming by the flapping of its valves, when it becomes a little more developed, the young Anodon attaches itself by means of its byssal filaments to either

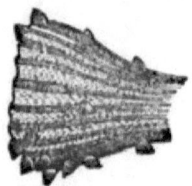

Fig. 25.—Young Mussels on the tail of a fish.

the gill-covers, lips, or fins, of a fish,—especially *Leuciscus* and *Gobio*,—and fixes its sharp teeth into its body. Remaining for a time in this parasitic condition, the single adductor muscle and byssal threads atrophy, and in their place the anterior and posterior adductors become developed, and the foot more developed. Changes go on until the larva has become like the parent from which it originated, and then the young Anodon loses its hold, drops down into the bottom of the water in which it exists, and commences the every-day life of its mother (*vide, Shell-Collector's Handbook for the Field*, p. 29). In *Sphærium* and *Pisidium* development takes place in a special brood-pouch developed near the inner gill-plate. The young bivalve is nourished by a secretion from the walls of this pouch, just as the young Anodon is nourished by a secretion from the gills. As was first recorded by Garner, it is not a rare thing to find the young of *Sphærium corneum* in the body of the adult animal.

Fig. 26.—A Group of Water Snails.

Paludina vivipara. *Limnaea stagnalis.* *Planorbis corneus.*

CHAPTER IV.

THE CLASSES, ORDERS, FAMILIES, GENERA, SPECIES, AND VARIETIES OF BRITISH LAND AND FRESH-WATER SHELLS.

(AQUATIC.)

CLASS I.—MALACOZOA ELATOBRANCHIA.

SHELL a bivalve, the two valves of which are united along their dorsal margin by a ligament. Body, oval, headless; mantle bi-lobed; foot tongue-shaped, sometimes provided with a byssus. Respiration performed by gills.

ORDER I.—LAMELLIBRANCHIATA.

Gills four in number, leaf-shaped, arranged in pairs on each side of the body between the visceral mass and the mantle

Family I.—Sphæriidæ.

Shell equivalve, subglobose; hinge with lateral and rdinal teeth. Body with one or two siphons at its anterior end.

1. **Sphærium.**—Shell nearly equilateral. Mantle with two prominent contracticle siphons.

(*a*) Shell yellowish-brown or brownish, finely striated, suborbicular; umbones blunt, nearly central; ligament small, not visible externally; muscular impressions faint; hinge strong with a double cardinal tooth in each valve, and two triangular-shaped teeth in the right valve and four teeth in the left. Length 6 lines; breadth 4 lines; thickness 3½ lines. Ditches, marshes, ponds, canals, and rivers. Generally distributed. *S. corneum.*[1]

---

[1] v. *flavescens*, paler, not so large, more globular; v. *nucleus*, smaller, nearly spherical; v. *Pisidiodes*, shell subtriangular, slightly more produced at its posterior slope, ligament just visible externally, transverse striæ coarser; v. *minor*, shell smaller, nearly globular.

(*b*) Shell brownish-green, or yellowish-horn colour, strongly striated, oval, ventricose; marked sometimes with two or three dark bands; umbones blunt, central; ligament conspicuous; muscular impressions distinct; teeth as in *S. corneum.* Length 10 lines; width 7 lines; breadth 5 lines. Canals and sluggish streams. *S. rivicola.*

(*c*) Shell pale grey, compressed, oblong, faintly striated; umbones small, central; ligament long, narrow, visible externally; cardinal teeth very small; muscular impressions distinct. Length ½ inch; width ⅜ths inch; thickness ¼ inch. Ponds and rivers. Local. *S. ovale.*

(*d*) Shell yellowish-white or ashy, more compressed than *S. corneum*, subrhomboidal; umbones narrow, central, capped with the nucleus of the shell; ligament narrow, slightly visible; muscular impressions indistinct, teeth smaller and shorter. Length 4 lines; width 3 lines; breadth 1½ lines. Ditches, marshes, ponds, canals, and rivers. Generally distributed, but local. *S. lacustre.*[2]

2. **Pisidium.** — Shell inequilateral. Animal with one siphon.

(*a*) Shell triangular, tumid, deeply grooved concentrically, whitish-grey or pale brown; umbones blunt, rather prominent; ligament short, narrow, conspicuous; teeth as in *Sphærium.* Length 4 lines; width 3 lines; thickness 1½ lines. Canals, ponds, lakes, and rivers. *P. amnicum.*[3]

(*b*) Shell triangular, very tumid so as to be cuboid in shape, greyish-white, finely striated; ligament short, scarcely visible; umbones pointed, prominent; muscular impressions distinct. Length 2¼ lines; width 1¾ lines. Canals, ponds, ditches, and sluggish streams. *P. fontinale.*[4]

(*c*) Shell oval, compressed but swollen, yellowish-white or greyish-horn colour, thin, finely striated; umbones blunt, short,

---

[2] v. *rotunda*, shell rounder, flatter, yellowish-green; v. *Ryckholtii*, shell small, globular, triangular, umbones pronounced; v. *Brochoniana*, shell larger, flatter, umbones smaller.

[3] v. *striolata*, smaller, wrinkles more pronounced; v. *læviuscula*, larger, wrinkles demi-effaced.

[4] v. *cinerea*, larger, flatter, striæ fainter; v. *pulchella*, more glossy, strongly grooved, umbones less acute; v. *Henslowana*, with a plate-like appendage near the umbones.

nearly central; ligament short, not visible from the outside. Length 1¾ lines; width 1½ lines. Marshes and ditches.

P. pusillum.[5]

(*d*) Shell suborbicular, in the upper part rather tumid, in all the other parts compressed, thin, very glossy; ligament short, not visible from the exterior; muscular impressions distinct. Length $\frac{1}{12}$th inch; width ⅓th less. Ponds and pools. *P. nitidum.*[6]

(*e*) Shell very small, oblong, thin, ventricose, deeply striated, yellowish-white or pale horn-coloured; umbones blunt, prominent, very excentric; ligament nearly invisible; muscular impressions indistinct. Ponds and pools. *P. roseum.*

Family II.—Unionidæ.

Shell large, oblong, equivalve, inequilateral. Mantle lobes free

Fig. 27.—*Unio Pictorum.*

all round, except at the posterior edge, where they form two orifices, which are somewhat awl-shaped.

[5] v. *ventrosa*, shell slightly more trigon-shaped, more ventricose, v. *grandis*, much larger, v. *obtusalis*, shell smaller, much more tumid, umbones sharp and pronounced.

[6] v. *splendens*, lemon-coloured.

3. **Unio.**—Shell firm, solid, cardinal teeth large, lateral teeth lamelliform.

(*a*) Shell oblong, ovate, solid, brown; epidermis smooth; umbones prominent, rugose, excentric; anterior teeth high, conical, strong. Length 3 inches; width ½ inch. Canals, ponds, rivers.

U. tumidus.[7]

(*b*) Shell oblong, ventricose, wedge-shaped, greenish-yellow marked with brown in lines of growth; ligament parallel to lower margin, longer than in *U. tumidus;* umbones not so prominent nor so rugose as in *U. tumidus;* anterior teeth arched, much compressed; hinder teeth of left valve small or evanescent. Length 2 to 3 inches. Ponds, lakes, and slow running streams.

U. Pictorum.[8]

(*c*) Shell black-brown, elongated, ovate, thick, heavy, strongly striated, very compressed; umbones incurved, generally eroded, excentric; cardinal teeth small, slightly compressed, thick, subtriangular, conical, the posterior one of the left valve much developed. Length 5¼ inches; width 2½ inches; breadth 1 inch. Among the gravel and small stones in the shallows of quick running rivers or mountain torrents. *U. margaritifer.*[9]

4. **Anodonta.**—Shell thin, hinge without teeth.

(*a*) Shell very large, oval, thin, dull green, more or less dusky; lines of growth large; umbones small, plaited; ligament long, parallel to lower margin; posterior side gaping and produced into a rounded cuneiform point. Length 2½ to 4 inches; breadth 3 to 9 inches. Lakes, ponds, canals, and muddy rivers.

A. cygnea.[10]

(*b*) Shell smaller than *A. cygnea*, oval, compressed behind,

---

[7] v. *radiata*, thinner, greenish with yellow rays, hinge-line nearly straight; v. *ovalis*, wedge-shaped, dark olive-brown.

[8] v. *radiata*, yellowish with green rays; v. *latior*, broader, shorter, yellowish-brown; v. *curvirostris*, smaller, shorter, bent, olive-coloured; v. *rostrata*, slightly more elongated, lanceolate behind, brownish.

[9] v. *Roissyi*, less arched above, not sinuous below, larger posteriorly; v. *sinuata*, broader in proportion to length than type, yellowish-brown, lower margin incurved towards the middle.

[10] v. *ventricosa*, shell larger, more elongated, more ventricose; v. *radiata*, larger with yellow rays; v. *Zellensis*, yellowish-brown, broader, upper and lower borders nearly parallel; v. *pallida*, light yellow or fawn colour; v.

brown-olive, glossy; umbones straight, excentric; umbonal region compressed, rugose; ligament short, prominent, forming an angle with the lower margin; upper margin curved and raised into a sort of a crest, length 2½ inches; breadth 3½ inches. Small rivers, canals, and ponds. *A. anatina.*[11]

Family III.—Dreissenidæ.

Shell boat-shaped, equivalve, furnished with a byssus; umbones placed at the extreme end; hinge with small teeth or edentulous ligament internal.

5. Dreissena.

(*a*) Shell mussel-shaped, triangular, keeled in the centre of both valves, olive or yellowish-brown, marked transversely with zigzag streaks of purple or dark brown; ligament long, narrow, and fitting into a groove in the hinge of both valves; umbones incurved, small, placed at the anterior end; lower margin incurved, length 1 to 1¾ inch. Canals, lakes, and rivers.

D. polymorpha.

Class II.—MALACOZOA GASTROPODA.

Shell univalve or none. Body with a distinct head, and two or four tentacles; eyes situated at the extremity of the dorsal tentacles or at the base of them; respiration effected by gills or lung.

Order I.— Pectinibranchiata.

Shell spiral, external, operculated. Respiratory organ consisting of a single pectiniform gill.

intermedia, oblong, feebly arched below, moderately arched above, obliquely truncated behind; v. *rostrata*, oblong-oval, upper margin forming a dorsal crest.

[11] v. *complanata*, oval, very compressed, brown, umbones close to anterior margin; v. *radiata*, marked with yellow and green rays; v. *ventricosa*, very tumid in the middle and umbonal region, larger, more solid, marked with green and yellow rays; v. *minima*, shell smaller, not quite so broad, blackish. (By some conchologists *Anodonta anatina* is considered to be a variety of *A. cygnea*, but a recent examination of over seven hundred specimens does not warrant me in doing likewise.)

Family I.—Neritidæ.

Shell semiglobose; spire small, flat, excentric, mouth semicircular; operculum shelly, with a plate-like appendage on its under side.

1. **Neritina.**

(*a*) Shell ovate, solid, glossy, yellowish or brownish, with white purple, brown or pink chequerings; spire short, lateral, composed of three whorls, the last one being excessively disproportionate; mouth semilunar; operculum semilunar, yellowish, bordered with orange, and with a strong, raised, grooved spire placed at one end of its lower surface. Length 3 lines. In canals and rivers with stony bottoms. *N. fluviatilis.*[12]

Family II.—Paludinidæ.

Shell cone-shaped, ventricose; mouth oval; operculum concentrically striated. Body oval; eyes sessile or placed on pedicels at the base of the tentacles; gills internal.

2. **Paludina.**—Animal ovoviviparous; eyes placed on pedicels; operculum horny.

(*a*) Shell conically oval, ventricose, dull yellowish-green with three brown bands on the body whorl, and two upon each of the preceding whorls; umbilicus none or represented by a slit; mouth an irregular oval. Length 1 to $1\frac{3}{5}$ths inch. Ponds, lakes, canals, and slow rivers. *P. vivipara.*[13]

(*b*) Shell conoid, very ventricose, brown olive marked with three brownish bands on the body whorl, and with two on the preceding whorls; spire blunt or sharp: sutures very deep; umbilicus distinct, narrow, deep; mouth nearly circular. Length 1 to $1\frac{3}{5}$ths inch. Canals and stagnant waters. *P. contecta.*[14]

3. **Bythinia.**—Animal oviparous; operculum shelly.

(*a*) Shell ovoid, somewhat conical, glossy, thin, solid, yellowish-horn colour, semi-transparent; whorls six, convex, the body-whorl

---

[12] v. *cerina*, straw-yellow, unicolor; v. *nigrescens*, black or blackish; v. *undulata*, with some transverse dark bands; v. *trifasciata*, with three spiral dark bands.

[14] v. *albida*, white; v. *unicolor*, bandless; v. *atro-purpura*, black, which when viewed by transmitted light is purple.

[14] v. *virescens*, greenish, unicolor.

large; mouth oval, angulated above; umbilicus none; operculum oval, angulated above, closely fitting. Length ½ inch. Streams, ditches, canals. *B. tentaculata.*[15]

(*b*) Shell conoid, swollen towards the base, thin, subtransparent, yellowish-horn coloured; whorls four to five, tumid, separated by a deep suture; mouth nearly circular, less angulated above than in *B. tentaculata;* umbilicus small, distinct; operculum nearly circular. Length ¼ inch. Ditches, canals, and sluggish rivers. More local than *B. tentaculata.* *B. Leachii.*[16]

4. **Hydrobia.**—Operculum horny, thin; eyes placed on tubercles.

(*a*) Shell ovoid, subopaque, thin, yellowish-horn coloured; whorls five to six, very convex; sutures deep and grooved; umbilicus represented by a narrow chink; mouth oval; operculum oval, nucleus lateral. In ditches between Greenwich and Woolwich which are flooded by the tide. *H. similis.*

(*b*) Shell a lengthened cone, thin, glossy, yellowish-horn coloured; whorls six to seven, convex; sutures not grooved; mouth oval; umbilicus smaller than in *H. similis.*

H. ventrosa.[17]

Family III.—Valvatidæ.

Shell conoid, more or less depressed; mouth circular; operculum horny. Body spiral, with two tentacles, provided with a long plume-like gill which is protruded when the animal is crawling; eyes situated on the inner side of the base of the tentacles.

5. **Valvata.**

(*a*) Shell globular, thin, solid, light horn-coloured, whorls five to six, rounded, body-whorl very large; spire obtuse, compressed;

---

[15] v. *albida*, white; v. *fulva*, tawny, glossy; v. *ventricosa*, more tumid, globular-conical in shape, white; v. *zonata*, reddish with one or more white lines or bands; v. *producta*, less tumid, in shape an elongated cone; v. *excavata*, whorls more rounded, suture much deeper; m. *decollatum*, with the apex decollated; v. *pallida*, shell very pale horn coloured; m. *scalariforme*, spire scalaria.

[16] v. *albida*, white; v. *elongata*, smaller, spire more produced.

[17] v. *pellucida*, clear white, nearly transparent; v. *minor*, smaller, spire shorter; m. *decollatum*, spire decollated.

mouth circular, with a complete peristome; umbilicus deep; operculum circular, greyish-white. Length ¼ inch. Lakes, ditches, canals, and rivers. *V. piscinalis.*[18]

(*b*) Shell flatly coiled pale horn coloured, glossy; whorls five, the last one very large; mouth circular with a continuous margin; umbilicus large and open, exposing the interior convolutions; operculum round, reddish-horn coloured. Diameter $\frac{1}{10}$th inch. Lakes, ponds, canals, and ditches. *V. cristata.*

Order II.—Pulmonobranchiata.

Shell generally spiral and external, but sometimes (*Limacidæ*) rudimentary and internal, or wanting. Body spiral, generally non-operculated, but sometimes with an operculum; respiration effected by means of a lung.

Family I.—Limnæidæ.

Shell spiral or hood-shaped; mouth without teeth. Tentacles two; eyes sessile.

6. **Planorbis.**—Shell orbicular, flat, coiled nearly in the same plane; mouth semi-circular; umbilicus distinct; non-operculated. Tentacles two in number, very long; eyes sessile; foot oval, short.

(*a*) Shell quoit-shaped, depressed, yellowish-horn coloured, glossy; whorls four with two to five curved transverse plates inside the last whorl, which appear as whitish lines when viewed from the exterior; periphery bluntly carinated; umbilicus narrow, deep. Diameter $\frac{1}{5}$th inch. Ponds and slow streams. Local.

P. lineatus.[18]

(*b*) Shell quoit-shaped, thin, glossy, yellowish-horn coloured, more or less reddish; whorls four to five, the outer whorl being without septa as in *P. lineatus*, and very large in proportion to the rest, covering one half of the preceding whorl, and bluntly keeled in the middle of its periphery; spire not so much sunk as in *P. lineatus;* umbilicus small, shallow. Diameter 2½ lines. Ponds and ditches on aquatic plants. *P. nitidus.*[19]

[18] v. *depressa*, more depresssed, umbilicus large; v. *albina*, white or whitish; v. *pusilla*, smaller, striæ, stronger whorls four and a half; v. *acuminata*, spire produced, apex sharp.

v. *albina*, milk-white, transparent.

[19] v. *albida*, white; v. *minor*, smaller.

(*c*) Shell quoit-shaped, thin, rather concave above, rather convex below, not glossy, dull light brown or grey, subtransparent; whorls three, the outer whorl bluntly and indistinctly keeled on its periphery, and strongly marked with transverse ridges in line of growth; mouth oval; umbilicus large. Diameter $\frac{1}{10}$th inch. Ponds and ditches on aquatic plants. *P. nautileus*.[20]

(*d*) Shell convex above, concave below, thin, whitish, transparent, finely striated longitudinally, and marked with fine close-set raised circular striæ which are clothed with deciduous bristles; whorls five; mouth roundish-oval; umbilicus large. Diameter ¼ inch. Lakes, ponds, and stagnant water. *P. albus*.[21]

(*e*) Shell convex above with a central depression, concave below, brownish-horn coloured, smooth, glossy; whorls five to six, more convex than in *P. albus*; mouth nearly circular; suture well defined; umbilicus large. Diameter 2 lines. Ponds, marshes, and lakes. Very local. *P. parvus*.[22] (= *P. glaber*.)

(*f*) Shell concave above, flat below, *or the reverse*, thick, horn-coloured; whorls five to six, rounder and with the keel not so well pronounced as in *P. albus*; mouth roundish; peristome often white ribbed; umbilicus wide, shallow. Diameter, ¼ inch. Shallow and stagnant waters, sluggish streams. Common.

P. spirorbis.[23]

(*g*) Shell very flat, concave above, flat below, thin, glossy, brownish-horn coloured, transparent; whorls six to eight, the outer whorl sharply keeled on its lower margin; mouth rhombic, compressed; umbilicus large, shallow. Diameter ⅜ths inch. Shallow and stagnant waters. Moderately common.

P. vortex.[24]

[20] v. *crista*, transverse ridges more pronounced, periphery deeply notched or crested by them.

[21] v. *Draparnaudi*, more closely striate in line of growth; periphery distinctly keeled; umbilicus larger. (This species is often found covered with dirt, and requires to be well cleaned in order to show its characters.)

[22] v. *compressa*, more concave below, depressed in the centre only on the upper side, whorls rounder and not increasing so quickly.

[23] v. *albida*, white; v. *ecarinata*, smaller, light green, keelless, one whorl less than usual.

[24] v. *compressa*, thinner, flatter, keel more distinct. and sharper, and placed nearly in the middle of periphery.

(*h*) Shell nearly flat above, rather convex below, yellowish-horn coloured, thin, glossy; whorls five to six, the outer whorl being sharply keeled in the middle line of its periphery; mouth obliquely oval, somewhat angular; umbilicus small. Diameter $\frac{1}{2}$ inch. Stagnant waters and sluggish rivers. Local. *P. carinatus.*[25]

(*i*) Shell slightly concave above, flattish, with a central concavity below, striolate, brown-horn coloured; whorls five to six, rapidly enlarging, with the periphery strongly keeled below; suture deep; mouth rhombic, rounded in front, often ribbed internally; umbilicus large, shallow. Diameter $\frac{3}{4}$ths inch Ponds, canals, slow rivers, and ditches. *P. complanatus.*[26]

(*j*) Shell reddish-brown, very tumid, glossy, nearly opaque; whorls five to six, rounded above and below; periphery not keeled; mouth semilunar; umbilicus broad, shallow. Diameter $\frac{1}{2}$ to 1 inch. Canals, streams, marshes, and ponds. Somewhat local. *P. corneus.*[27]

(*k*) Shell nearly flat above with a deep concavity in the middle, very convex below, solid, opaque, yellowish or brown-horn colour; whorls eight, compressed; suture deep; mouth crescent shaped, very narrow; umbilicus large, deep. Diameter $\frac{2}{10}$ths inch. Ponds, ditches, and lakes. *P. contortus.*[28]

(*l*) Shell flattish, dull, very convex below; whorls two to two and a half, with the periphery of the outer whorl angulated; mouth large, squarish, and much expanded; umbilicus abruptly contracted, small, deep. Diameter $\frac{1}{10}$th inch. In canals round Manchester and Burnley. Imported. *P. dilatatus.*

7. **Physa.**—Shell spiral, thin, polished; spire *sinistral*, non-operculated. Animal with two long tentacles, with the eyes at their base; mantle very large so as to cover part of the shell.

(*a*) Shell fusiform, thin, yellowish or reddish-horn coloured, glossy, semi-transparent; whorls six to seven, rounded; suture

---

[25] v. *albida*, pellucid white; v. *disciformis*, yellowish, flatter, thinner, keel more prominent and sharp.

[26] v. *albina*, whitish or colourless; m. *terebrum*, whorls dislocated from one another, and elevated into a spiral cone; v. *rhombea*, smaller, more convex above, with a deep concavity below, keel blunt.

[27] v. *albina*, perfectly white.

[28] v. *albida*, nearly white; v. *excavata*, much depressed, sunken above.

shallow; mouth oval-lanceolate. Length ½ to ¾ths inch. Ponds, ditches, and slow-running streams. More or less local.

P. (Aplecta) hypnorum.[29]

(b) Shell oval, thin, glossy, semi-transparent, pale greyish-horn coloured; whorls four to five, tumid, the last one occupying ¾ths or ⅘ths of the shell; suture deep; mouth oblong and wider and larger in proportion than in P. hypnorum. Length ½ inch. Running brooks, canals, ditches, and sluggish rivers. Common.

P. fontinalis.[30]

(c) Shell an elongated ovoid, ventricose, glossy, finely striated longitudinally, light horny or whitish in colour; whorls three, the last one ⅗ths of the total height of the shell; suture moderately deep; apex sharp; mouth obliquely and narrowly oval, acute above. Height ⅓rd to ⅔rds inch. In one of the lily tanks in Kew Gardens. Imported. P. acuta.

8. **Limnæa.**—Shell oval, thin, translucent; mouth oblong; columella with an oblique plait; non-operculated. Animal with two short triangular tentacles bearing the eyes at their base; foot oval.

(a) Shell oval, globular, amber or yellowish-horn coloured, thin, transparent; whorls three to four, inflated, the last one forming nearly the whole of the shell; spire slightly produced; suture rather deep; mouth oval. Mantle partly covering the shell. Length ½ inch. Ditches, ponds, and lakes. Very local.

L. glutinosa.[31]

(b) Shell ovate, thin, fragile, glossy, pale amber-coloured, glossy, whorls three to four, convex, the body whorl very large; spire sunk within the last whorl; suture shallow; mouth pyriform, large. Length 5½ lines. In a small alpine lake on Cromaglaun Mountain, Killarney, Ireland. L. involuta.

(c) Shell oval, last whorl swollen, thin, glossy, yellowish-horn

---

[29] v. *cuprella*, copper-coloured, whorls six, convex, the last one far exceeding in size the rest put together, mouth oval, contracted above; v. *major*, larger, more coloured, height 10 to 13 mm; m. *decollatum*, spire decollated.

[30] v. *albina*, milk-white; v. *oblonga*, spire considerably produced; v. *inflata*, larger, more ventricose; v. *curta*, spire extremely short; m. *dextrorsum*, spire dextral.

[31] v. *mucronata*, not quite so globular, spire more produced.

colour; whorls five, convex, the body-whorl very large; spire raised and sharp at its apex; mouth large, oval, more than half the length of the shell. Length ½ to 1 inch. Stagnant waters and slow running streams. Common. L. peregra.[2]

(*d*) Shell globosely ovate, glossy, yellowish-horn coloured, semi-transparent; whorls four to five, the last one very much swollen and occupying ⅝ths of the shell; spire short, acute; mouth roundish oval, vastly expanded, oblique. Length 1 inch. Lakes, ponds, canals, marshes, and sluggish rivers. Moderately local.

L. auricularia.[33]

(*e*) Shell elongated, ovate, oblong, thin, greyish-white, horn or brown-coloured; whorls six to eight, striated in line of growth,

Fig. 28.—*Limnæa stagnalis.*

[2] v. *labiosa*, smaller, outer lip remarkably expanded and reflected; v. *ovata*, ampullaceous, glossy, whorls very convex, spire acute and very short, suture deep, aperture obliquely produced, ⅔ths the length of the shell; v. *acuminata*, resembling v. *ovata* but has the spire more produced and the mouth smaller; v. *Burnetti*, spire exceedingly short, nearly truncate, intorted; v. *nitida*, larger, slightly transparent, fauve, coloured; v. *solemia*, ventricose, whorls rounded, spire short, fauve, subtransparent; v. *succineæformis*, shaped like a succinea, whorls four; v. *intermedia*, rather compressed towards front margin, thinner, spire more produced, mouth expanded.

[33] v. *albida*, smaller, thinner, white, spire shorter, striæ less distinct; v. *reflexa*, outer lip much reflected; v. *magna*, larger, aperture narrower, outer

the body-whorl occupying nearly ¾ths the length of the shell; spire acute, tapering; mouth oval, large. Length 1½ to 2 inches. In stagnant and slow-running waters. Common. *L. stagnalis.*[34]

(*f*) Shell conical, tapering, thick, opaque, dull brown; whorls six to seven, convex, the last whorl occupying ⅔rds the length of the shell; spire produced, acute; suture rather deep, circled by a narrow white line; mouth ovate, outer lip marked with brown or violet. Length ¼ to ¾ths inch. Ponds, marshes, and lakes. Common. *L. palustris.*[35]

(*g*) Shell oblong-oval, turreted, pale brown or yellowish-horn coloured, rather glossy; whorls five to six, convex, deeply separated from one another; spire produced; apex acute; suture very deep; mouth ovate, oblong; umbilical cleft distinct. Somewhat resembles *L. palustris.* Length ½ inch. Marshes, ditches, pools, and muddy streams. Common. *L. truncatula.*[36]

(*h*) Shell tapering, an elongated cone, thin, glossy, horn-coloured or brownish; whorls seven to eight; mouth elongate, ovate, not above ⅓rd the length of the shell, provided with a white internal rib; umbilical cleft very minute. Length 1 inch. Ponds and ditches. Local. *L. glabra.*[37]

---

margin nearly parallel to the columella which is straight, the upper edge reaching the commencement of the spire which is sharp; v. *acuta*, smaller, more oblong, having the last whorl and mouth proportionably narrower.

[34] v. *fragilis*, smaller, narrower, thinner, amber-coloured; v. *labiata*, dwarfed, with the outer lip thickened and reflected; v. *roseolabiata*, slightly narrower, brown-black, mouth bordered interiorly with rose-violet; v. *elegantula*, nearly scalariform, suture deep.

[35] v. *corva*, larger, swollen, opaque, blackish, violet-coloured within; v. *albida*, white; v. *obesa*, very tumid; v. *conica*, conic, greyish-white, suture deep, umbilical cleft present; v. *tincta*, shorter, broader, light-brown, mouth purplish; v. *lacunosa*, with flattenings, depressions, and protuberances; v. *elongata*, slightly larger, slightly narrower, opaque, brown, spire much produced; m. *decollatum*, spire decollated.

[36] v. *albida*, smaller, milk-white; v. *major*, larger, ashy, more tumid; v. *ventricosa*, more bellied, spire short, peristome without swelling; v. *minor*, smaller, horn-coloured; v. *elegans*, much larger, more solid, more slender, greyish-white, marked with coarse spiral ridges, spire more produced, suture oblique, outer lip thickened.

[37] v. *major*, much larger; v. *elongata*, more produced, so as to alter the relative proportions of length and breadth; m. *decollatum*, spire decollated.

9. **Ancylus.**—Shell conical, oblong, limpet-shaped; apex pointed and bent to the right; spire dextral or sinistral. Animal with two cylindrical tentacles, with the eyes at their base; foot large.

(*a*) Shell dextral, conoid, thin, yellowish-grey or horn-coloured; apex recurved; mouth ovate. Height $\frac{1}{4}$ to $\frac{1}{8}$th inch. Streams and rivulets. Common as far north as Aberdeen. *A. fluviatilis.*[38]

(*b*) Shell sinistral, oblong, horn-colour tinged with yellow or green, thin, glossy, compressed at the sides; apex acute, subcentral; mouth somewhat oblong. Length $\frac{1}{4}$ inch. Lakes, ponds, canals, and sluggish streams. Somewhat local.

A. lacustris.[39]

(TERRESTRIAL.)

Family II.—Limacidæ.

Shell placed under the mantle, granular or shield-like. Body united in its whole length with the foot beneath, there being no visceral hump; tentacles four, cylindrical, the dorsal pair bearing the eyes; mantle shield-like.

10. **Arion.**—Shell consisting of calcareous granules. Mantle

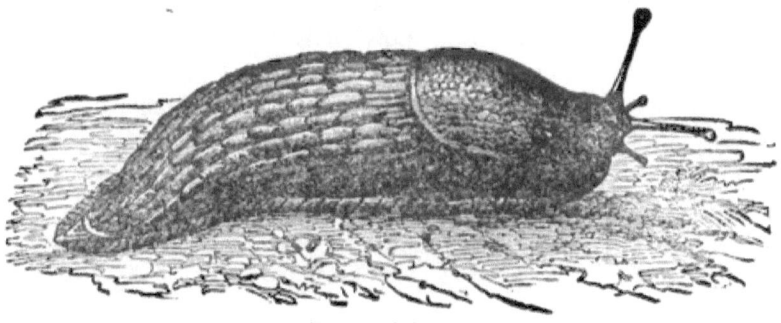

Fig. 29.—*Arion ater.*

[38] v. *capuloides*, elevated, very convex anteriorily and laterally, nearly convex behind, aperture oboval; v. *albida*, milk-white, striæ finer; v. *gibbosa*, slightly elevated, extremely convex, subgibbous in front, convex laterally, nearly straight posteriorly, apex obtuse, aperture oboval; v. *costata*, whitish or greyish, striæ well marked.

[39] v. *albida*, milk-white; v. *compressa*, rather larger and considerably broader and flatter.

shagreened, not striated concentrically; respiratory orifice placed on anterior half of mantle; tail provided with a slime gland.

(*a*) Body varying in colour, coarsely tuberculated, rounded in front, attenuated behind; tentacles black, bulbs much swollen; margin of foot marked with transverse black lines; slime yellowish. Shell none, or consisting of loose granules. Length 2 to 5 inches. Type black. Woods, gardens, and hedge banks. Common. *A. ater.*[40]

(*b*) Animal smaller than *A. ater*, marked with longitudinal grey bands; mantle with a band down its middle and one round its margins; margin of foot red, orange, or yellow; slime yellowish. Shell consisting of calcareous granules cemented into an oval mass. Length 1 to 1½ inches. Woods, hedges, gardens. Common. *A. hortensis.*[41]

(*c*) Animal cylindrical, reddish-brown, with two lateral faint blackish bands; mantle very convex, paler than back, with two lateral black bands; tentacles blackish; back somewhat keeled in its hinder portion and finely scaled; pulmonary aperture nearly median; foot grey, marked with small transverse black lines; slime saffron-yellow. Shell a few loose granules. Length 1½ to 2½ inches. Woods and damp places. *A. subfuscus.*

(*d*) Animal whitish-grey; back blackish with lateral bands; keel well pronounced in young individuals, but becoming more and more obsolete as the slug becomes adult, and represented then by a pale line; foot whitish. Length 1½ inches. Woods, gardens, fields. Common. *A. Bourguignati.*

11. **Geomalacus.**—Shell solid, unguiform, concentrically striated. Body capable of great extension; mantle finely shagreened; tentacles short, eyeless; respiratory orifice placed more anteriorily than in *Limax*; tail provided with a large slime gland.

(*a*) Animal black, spotted with yellow, coarsely tuberculated; foot brown, transversely furrowed; sole light yellow divided into

---

[40] v. *rufa*, red, unicolor; v. *brunnea*, coffee or rust coloured; v. *Draparnaudi*, dark-red, foot-fringe yellowish or reddish; v. *succinea*, yellowish, unicolor; v. *marginata*, black, foot-fringe yellow, orange, or lead-coloured; v. *pallescens*, dirty white.

[41] v. *grisea*, pale grey, unicolor.

three bands by a median longitudinal band.   Length 2 inches.
West Kerry, Ireland. *G. maculosus.*

12. **Amalia.**—Shell oval or elliptical. Mantle granulated, having the respiratory orifice on its posterior half; back keeled from tail to mantle; slime-gland absent.

(*a*) Animal varying in colour; mantle oblong, bilobed, rounded behind, somewhat truncated anteriorily; tentacles slate-coloured; slime colourless, glutinous. Shell oval, rugous, thick from nucleus to centre. Length 2½ inches. Gardens and hedgerows. Local. *A. gagates.*[42]

(*b*) Animal varying in colour, truncated in front; mantle an elongated oval, not bilobed; head and tentacles dusky; keel of a lighter colour than the rest of the back; slime colourless, glutinous. Shell elliptical, somewhat concave, very thick. Length 2½ inches. Gardens and fields. Common. *A. marginata.*[43]

13. **Limax.**—Shell oval, oblong, quadrangular, or unguiform. Mantle concentrically striated; respiratory orifice situated on posterior half of mantle; tail keeled, not the back; slime gland absent.

(*a*) Animal fleshy, yellow, tessellated with black and white, coarsely tuberculated; mantle broadly rounded behind; head, neck, and tentacles slate colour; sole milk-white; slime yellow. Shell oblong or quadrangular, thin, concave, margin membranaceous. Length, 4 inches. Cellars and damp places. Common. *L. flavus.*[44]

(*b*) Animal varying in colour, generally creamy-white mottled with dusky grey; mantle large, broadly rounded behind; foot pale grey or cream colour; slime milk-white. Shell small, oval, thin, margin membranaceous. Length ¼ to 1½ inches. Fields, gardens, woods. Very common. *L. agrestis.*[45]

---

[42] v. *plumbens*, lead-coloured; v. *rava*, drab-coloured; v. *olivacea*, olive-coloured.

[43] v. *rustica*, greyish, mantle reddish, with a longitudinal black band on each side; v. *rufula*, yellowish-red.

[44] v. *virescens*, greenish, spots indistinct; v. *grisea*, ground-colour grey instead of yellow, otherwise like the type; v. *suffusa*, grey, unicolor.

[45] v. *filans*, greyish-white or ashy, mantle yellowish; v. *nigra*, jet black, tentacles bluish or brownish-black, sole pale; v. *sylvatica*, greyish, mottled;

(c) Animal dark brown, slender, very glossy; mantle very tumid behind, pale yellowish-brown; sole ash-coloured; slime almost colourless. Shell unguiform, convex above, flat beneath, solid; margin not membranaceous. Length ½ to ¾ths inch. Very active. In damp situations. Moderately common. *L. lævis.*

(d) Body rounded, compressed towards the tail, greenish-white, slender; head and tentacles black; mantle yellowish; slime thick, orange-coloured. Shell oval, somewhat tuberculous, margin broad, thin, membranaceous. Length ⅞ths inch. Woods in Shetland and Northumberland. Local. *L. tenellus.*

(e) Body varies much in colour, banded; mantle pointed behind; dorsal tentacles short; foot margin white; slime colourless. Shell oval, thin, nearly flat, margin broad and membranous. Length 1½ to 3 inches. In woods on trees, especially the beech and walnut. Somewhat local. *L. (Lehmannia) arborum.*[46]

(f) Body large; tentacles long, vinous-coloured; mantle buckler-shaped, swollen, produced behind; slime whitish. Shell oblong, solid, convex above, concave below, margin membranaceous. Length, 3 to 6 inches. Cellars, gardens, woods, fields, hedgerows. Common. *L. maximus.*[47]

(g) Mantle unicolorous; pulmonary orifice margined with the same colour as the ground colour of the body but of a deeper tint; keel different in colour from the rest of the body; sole divided

v. *punctata*, greyish-white, minutely spotted with black; v. *tristis*, brownish, mantle with two lateral brown bands, and sometimes a third intermediate band; v. *cineracea*, greyish-white, mantle ashy (type); v. *submaculata*, greyish-white, streaked with seal-brown on the back which extends on to the posterior ⅔rds. of the mantle, the sides of the body, and anterior ⅓rd of the mantle are free from streaks and spotted with black.

[46] v. *maculata*, cinereous with the markings reduced to small and sharply defined black spots of a rounded or elongated form, and with a continuous band on each side showing a tendency to break into spots; v. *dicipiens*, brownish-grey, markings coalesced so as to produce the appearance of pale spots on a dark grey ground, lateral bands on mantle ill-defined, no bands on body, keel short, dorsal line partly obsolete.

[47] v. *obscura*, brown, unicolor; v. *maculata*, ashy, mantle and back spotted irregularly with black; v. *cinerea*, ashy, spotless, mantle bluish-black; v. *Johnstoni*, ashy, mantle spotted with black, back marked with spots and two bands of the same colour; v. *rufescens*, reddish, unicolor; v. *fasciata*, deep ash colour with whitish bands, often five in number.

into three longitudinal bands, the middle one being white, the two lateral ones dark. Length 3 to 6 inches. Gardens and fields. Local. *L. cinereo-niger.*

Family III.—Testacellidæ.

Shell small, auriculate, external, placed on the hinder portion of the body, covering the mantle; respiratory orifice on the right side below the mantle.

14. **Testacella.**

(*a*) Shell roundish-oval, depressed; whorls one and a-half, the nucleus making an angle of 45° to the vertical line; spire short; suture deepish; mouth roundish, dilated in front. Length ¼ to ⅖ths inch. Animal with two longitudinal furrows on its back, which commence from the shell and terminate near the head. Market gardens and fields. Local. *T. haliotidea.*

(*b*) Shell resembling that of *T. haliotidea* but smaller in proportion to its length, flatter on its superior surface, comparatively longer, more wedge-shaped, brownish, lines of growth finer and less rugged, thinner, and the left side of the shell is more strongly curved. Animal slightly more attenuated in front, and the longitudinal furrows near the mantle are much closer together. The nucleus of the shell is placed at an angle of 60° or 70° to the vertical line. The vas deferens enters the penis "terminally in *T. scutulum* and *Maugei:* laterally in *T. haliotidea.*" Gardens and fields. Local. *T. scutulum.*

Fig 30.—A snail-slug (*Testacella*).

(*c*) Shell resembling that of *T. haliotidea*, but larger and more cylindrical. Body dark brown. Gardens and fields round Bristol. *T. Maugei.*

Family. IV.—Helicidæ.

Shell spiral. Body distinct from the foot, tentacles four, retractile, cylindrical, the upper pair being the longest and bearing the eyes at their apices.

15. **Succinea.**—Shell oval or oblong, thin, transparent; spire short; mouth large and obliquely oval; non-operculated. Animal not capable of entirely entering its shell.

(*a*) Shell, ovate, oblong, smooth, glossy, amber-coloured, trans-

Fig 31.—*Succinea putris.*

parent; whorls three to four, ventricose, the body-whorl occupying $\frac{4}{5}$ths the length of the shell; mouth ovate, two-thirds the length of the shell. Length $\frac{1}{2}$ to $\frac{3}{4}$ths inch. On the banks of ditches and streams on flags and willows. Moderately common.

S. putris.[48]

(*b*) Shell oblong, suboval, thin, greenish-yellow, transparent; spire very short; mouth ovate. Length $\frac{2}{5}$ths inch. Marshes and ditch-banks. Moderately rare.

S. virescens. (= *Succinea putris* var. *vitrea*.)

(*c*) Shell amber-coloured, semi-transparent; compared with that of *S. putris* it is smaller, more slender, with a longer and more pointed spire; the suture near the mouth is also much deeper. Length $\frac{1}{2}$ inch. Marshes and ditch-banks. Moderately common.

S. elegans.[49]

(*d*) Shell oval, elongated, pale yellow or reddish, transparent; whorls three, twisted, the body-whorl occupying near the whole of the shell; spire short, with the apex tuberculous; suture oblique, well defined but not deep; mouth an elongated oval, about $\frac{2}{3}$rds

---

[48] v. *subglobosa*, shorter, broader, smaller, more solid; v. *solidula*, reddish-yellow, much thicker; v. *Ferussina*, deep reddish-fawn coloured, small, slightly elongated.

[49] v. *albida*, white; v. *minor*, thinner, reddish-brown, spire shorter, aperture more expanded; v. *ochracea*, thicker, smaller, spire longer, aperture smaller.

of the total height of the shell. Length $\frac{12}{25}$ths inch. Marshes and ditch-banks. Moderately common. *S. Pfeifferi.*[50]

(*e*) Shell small, oval, somewhat like that of *L. truncatula*, light horn-coloured; whorls four, convex; suture deepish; spire produced, with a blunt apex; mouth oval, about the same length as the spire. Distinguished from *L. truncatula* by having no reflected lip on the columella. Length $\frac{1}{4}$ inch. Edges of ditches near the coast, and on sanddunes near the sea. Rare. *S. oblonga.*

16. **Vitrina.**—Shell subglobular, thin, flattened; mouth large and semilunar; umbilicus wanting.

(*a*) Shell depressed, thin, glossy, glassy-green; whorls three to four, the body-whorl being very large; spire very short; mouth large, somewhat oval. Height $\frac{1}{12}$th inch. Among moss and dead leaves in woods and hedge-banks. Moderately common.

V. pellucida.[51]

7. **Hyalina. (Zonites).**—Shell orbicular, depressed, umbilicated; mouth obliquely crescent-shaped.

(*a*) Shell slightly convex above, glossy, striæ demi-effaced, reddish above, whitish below, especially round the umbilicus; whorls six to seven, the last whorl slightly dilated towards the aperture; umbilicus largish; mouth oval, very oblique. Diameter $\frac{1}{2}$ to $\frac{3}{4}$ths inch. Very local. Found only, as yet, at Guernsey, Torquay, Falmouth, Bristol and Isleworth, London. *H. Draparnaldi.*

(*b*) Shell dirty yellow or pale horn-coloured, glossy, under surface white, especially about the umbilicus; whorls five to six; spire very flat; umbilicus rather open; mouth crescent-shaped. Diameter $\frac{1}{2}$ inch. Under stones in fields and woods, about walls in gardens. Common. *H. cellaria.*[52]

---

[50] v. *ventricosa*, spire more elevated, body whorl convex, aperture large, less elevated, a little more than half the total height of the shell; v. *propinqua*, shell elongated, ventricose, body-whorl very large, spire short, aperture large, oval; v. *elata*, shell very slender, spire long and very tortuous, body-whorl very contracted at its origin, aperture oval and scarcely one half of the total height of the shell; v. *virescens*, greenish, spire elongated, suture deep.

[51] v. *Dillwynii*, nearly globular, last whorl very convex, spire more prominent; v. *depressiuscula*, rather oval, flatter on both sides, spire scarcely raised.

[52] v. *complanata*, smaller, spire flatter; v. *compacta*, body-whorl less swollen, not so white beneath, more compact and convex.

(c) Animal emitting an alliaceous smell. Shell amber or horn-coloured, thin, upper surface smooth and darker than the lower surface, which is not so white as in the other species; whorls five, the last not so large in proportion as in *H. cellaria;* spire more raised than in *H. cellaria;* umbilicus moderately large, exposing the second whorl; mouth narrow, crescent-shaped. Diameter $\frac{1}{4}$ inch. Walls, gardens, woods. Slightly more local than *H. cellaria*.
H. alliaria.[53]

(d) Shell convex above, less convex below and only very slightly marked with white, darkish horn-coloured; whorls five to five and a half, the body-whorl occupying about one-half of the shell; spire slightly produced, umbilicus narrow, deep, not disclosing the penultimate whorl, mouth forming three-fourths of a circle. Diameter $\frac{3}{8}$ths inch. Woods and fields. Local.
H. glabra.

(e) Shell subpellucid, dull waxy-coloured above, paler below, clouded with opaque white round the umbilicus; whorls four to five, marked with irregular placed striæ, which are interrupted by the sutures and not continued from whorl to whorl; spire slightly produced; mouth semilunar; umbilicus very large. Diameter $\frac{3}{8}$ths to $\frac{3}{10}$ths inch. Under stones and amongst moss on hedgerows and old walls. Common. *H. nitidula.*[54]

(f) Shell whitish or light horn-coloured, glossy, thin, with numerous transverse circular striæ; whorls four, the body-whorl occupying nearly one-half of the shell; spire not much raised; suture deep, narrow; mouth roundish, oblique; umbilicus narrow, deep, disclosing all the internal spire. Diameter $\frac{1}{10}$th to $\frac{1}{8}$th inch. Under stones and decayed leaves in woods and fields. Somewhat local. *H. pura.*[55]

(g) Somewhat like *H. pura* but more glossy, regularly striated, horn-colour; whorls four to four and a-half with the striæ extending from whorl to whorl and not interrupted by the sutures as in *H. nitidula;* spire much depressed; mouth semilunar; diameter

---

[53] v. *viridula*, greenish-white.

[54] v. *Helmii*, pearl-white; v. *nitens*, much smaller, lighter coloured, las whorl expanded.

[55] v. *margaritacea*, pearl white, nearly transparent.

$1\frac{1}{2}$ths to $\frac{1}{8}$th inch; umbilicus moderately deep. Among moss and wet grass. Generally distributed, but not common.

H. radiatula.[56]

(*h*) Shell subglobular, chocolate-brown, no white on under side, solid, glossy, semi-transparent; whorls five, body-whorl occupying one-half of the shell; spire produced, with the apex blunt; mouth roundish, forming three-fourths of a circle; umbilicus narrow, deep. Diameter $\frac{1}{4}$ inch. Marshes and banks of ditches.

H. nitida.[57]

(*i*) Shell suborbicular, not very glossy, darkish horn-coloured, coarsely striated; whorls five and a-half; spire slightly produced; mouth nearly circular; umbilicus large, disclosing all the internal spire. Among moss and dead leaves in woods. Local.

H. excavata.

(*j*) Shell flat above, convex below, vitreous, smooth, glossy, transparent; whorls four and a-half to five, the body-whorl being scarcely smaller than the preceding whorl; mouth lunate; umbilicus very small. Diameter $\frac{4}{25}$ths inch. Among moss and decaying leaves in woods and hedgerows. Common.

H. crystallina.[58]

(*k*) Shell conical, dark horn-coloured, smooth, glossy, semi-transparent; whorls five and a-half to six; periphery bluntly keeled; spire produced; mouth semilunar, narrow; umbilicus shallow, indistinct. Diameter $\frac{1}{10}$th inch. Under stones and among decaying leaves in woods and hedgerows. Moderately common.

H. fulva.[59]

18. **Helix.**—Shell globular, turreted, convex or flattened; mouth more or less circular or oval; outer lip generally thick and possessing an internal rib, sometimes reflected and provided with tubercles or teeth; umbilicus sometimes absent but usually distinct.

(*a*) Shell conoid, globose, thin, horn-colour; epidermis raised into numerous close-set plaits in line of growth; whorls six, convex; spire blunt, slightly depressed; mouth crescent-shaped;

---

[56] v. *viridescenti-alba,* greenish-white.

[57] v. *albinos,* whitish.

[58] v. *complanata,* flatter on both sides, last whorl proportionably larger.

[59] v. *Alderi,* smaller, darker brown; v. *Mortoni,* depressed above, less flattened below, peripheral keel sharper.

umbilicus small, very deep, distinct. Height $\frac{2}{25}$ths inch. Among holly leaves in woods and groves. Local. *H. lamellata.*

(*b*) Shell conical, turreted, globose, brownish horn-colour; epidermis raised into numerous plaits, which, in the middle of each whorl, become produced into points; whorls four to four and a-half; aperture nearly crescent-shaped; umbilicus small. Height $\frac{3}{25}$ths inch. Among moss and dead leaves in woods and hedgerows. Moderately common. *H. aculeata.*[60]

(*c*) Shell very large, globose, coarsely striated, strong, whitish-yellow banded with brown; whorls four to five, the body-whorl very large; spire short, blunt; mouth roundish; inner lip reflected; umbilicus narrow. Height $1\frac{1}{25}$ths inch. Woods and hedgerows in chalky districts. Local. *H. pomatia.*[61]

(*d*) Shell conoid, globose, wrinkled, yellowish-brown with interrupted dark bands, solid; whorls four to five and a-half,

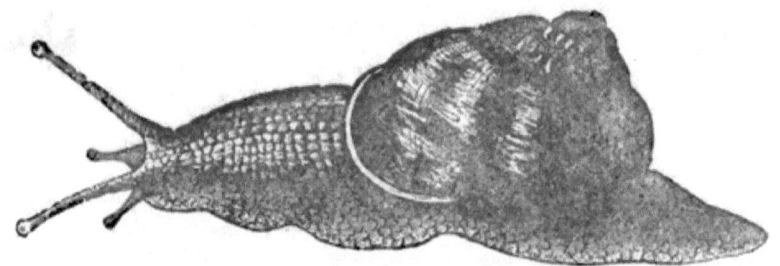

Fig. 32.—*Helix aspersa.*

convex, the body-whorl being very large in proportion; spire short, blunt; aperture oval; umbilicus none. Height $1\frac{7}{25}$ths inch. Gardens, fields, hedgerows. Very common. *H. aspersa.*[62]

(*e*) Shell subglobose, glossy, banded or unicolor; whorls five to five and a-half, convex; spire short, blunt at apex; mouth

---

[60] v. *albida*, whitish.

[61] v. *albida*, whitish, unicolor.

[62] v. *exalbida*, yellowish or whitish, unicolor; v. *grisea*, tawny or greyish, with extremely pale bands; v. *flammea*, reddish with longitudinal flame-shaped markings; v. *tenuior*, smaller, very thin, transparent, reddish, unicolor; v. *albofasciata*, reddish-brown, with a single white band; v. *globosa*, larger, nearly globular; v. *minor*, shell smaller; v. *conoidea*, thinner, smaller, conical, mouth small.

semilunar; inner lip thin, reddish-brown or chocolate-coloured; outer lip strong, reddish-brown; umbilicus covered by the inner lip. Height ⅜ths inch. Among nettles in woods and hedgebanks. Common. *H. nemoralis.*[63]

(*f*) Shell resembling that of *H. nemoralis* but smaller, more globular, rather thinner, and possessing a white or rose-coloured peristome. Same localities as last species. Common.

H. hortensis.[64]

(*g*) Shell globose, brown marked with yellowish spots, and a single black band round the periphery of each whorl; whorls five to six, convex; spire conoid, short, blunt; mouth rounded, lunate; peristome white; umbilicus distinct. Height $\frac{16}{28}$ths inch. Among the reeds and willows of ditch sides and river banks; in woods and hedgerows. Rather local.

H. arbustorum.[65]

(*h*) Shell somewhat depressed, thin, yellowish-horn coloured or creamy-white tinged with rose near the mouth, faintly banded with white on the periphery; whorls six to seven, convex, the body-whorl occupying one-half of the shell; spire short, blunt; mouth lunate, with a white or rose-coloured rib; umbilicus narrow, deep. Height $1\frac{1}{23}$ths inch. On nettles, and in marshy places, in some of the southern and northern counties of England. Local.

H. cantiana.[66]

[63] v. *castanea*, brown; v. *carnea*, flesh-coloured; v. *libellula*, yellow; v. *rubella*, pink; v. *albescens*, whitish; v. *hyalozonata*, with transparent bands; v. *major*, very large; v. *albolabiata*, with mouth and lip white; v. *bimarginata*, peristome brown exteriorly, white interiorly; v. *conoidea*, spire an elevated cone.

[64] v. *albina*, white or whitish; v. *pallida*, very pale yellow; v. *incarnata*, rose-coloured; v. *olivacea*, deep olive-brown; v. *lutea*, yellow; v. *minor*, dwarfed; v. *lilacina*, bluish-violet; v. *arenicola*, with transparent, colourless bands; v, *fuscolabiata*, peristome and rib dark; v. *roseo-labiata* rib pink or rose-coloured.

[65] v. *flavescens*, yellowish, unicolor; v. *major*, larger, spire more depressed; v. *conoidea*, conical, large, with transverse, somewhat confluent pale yellow markings; v. *alpestris*, half the usual size, more depressed; v. *fusca*, brown, nearly unicolor, thin, subtransparent.

[66] v. *albida*, white; v. *pyramidata*, smaller, spire more raised, pyramidal; v. *Galloprovincialis*, last whorl less depressed, lighter in colour, umbilicus more narrow, peristome reddish exteriorly, white interiorly.

(*i*) Shell depressed, subconic, not so glossy as *H. cantiana*, whitish-horn colour, with a white spiral band just above the periphery; whorls six to seven, faintly keeled, the body whorl very large; spire acute, more depressed than in *H. cantiana*; suture rather deep; mouth crescent-shaped with a white internal rib; umbilicus very small. Height $\frac{8}{25}$ths inch. Among grass on the Kent and Sussex downs, near the sea. Local.

H. Cartusiana.[67]

(*j*) Shell depressed, subconic, dull brown, more or less reddish, subcarinated, generally with a white band at the periphery; spire short, blunt; mouth semi-elliptical with a white internal rib; umbilicus narrow, deep. Height $\frac{6}{25}$ths inch; diameter $\frac{2}{5}$ths to $1\frac{1}{25}$ths inch. Under stones in hedge-banks. Common.

H. rufescens.[68]

(*k*) Shell suborbiculate, depressed, thin, yellowish-brown, epidermis covered with fine white recurved hairs arranged in spiral lines; whorls six to seven, convex, the body-whorls occupying one-third of the shell; spire blunt, slightly raised; aperture crescent-shaped; umbilicus narrow, deep. Height $\frac{1}{5}$th inch. Among moss, and under stones, etc., in shady places. Common.

H. hispida.[69]

(*l*) Like *H. hispida* but larger, less globose, hairs more liable to fall off; umbilicus wider. Height $\frac{1}{4}$ inch. In similar situations to the last. Common. *H. concinna*.[70]

(*m*) Shell subglobular, thin, dark brown or pale grey, marked with a few faintly-marked brown streaks, epidermis covered with white hairs; whorls six, convex, the body-whorl occupying nearly

---

[67] v. *leucoloma*, small, peristome and rib white; *rufilabris*, smaller, inside rib reddish.

[68] v. *alba*, entirely white; v. *rubens*, more or less reddish; v. *minor*, smaller, spire more raised.

[69] v. *fusca*, light brown; v. *subrufa*, reddish, glabrous, thicker; v. *nana* smaller, labial rib strong, spire depressed; v. *subglobosa*, more globular, thinner, horn-coloured or white, umbilicus very small; v. *conica*, smaller, spire more raised; v. *albida*, thin, white or colourless.

[70] v. *minor*, smaller, spire more raised; v. *alba*, white. (This species is probably a variety of *H. hispida*, as structurally it is not distinct.)

one-half ot the shell; spire blunt, much raised; mouth crescent shaped, with a white internal rib; umbilicus deep, very minute. Height $\frac{9}{25}$ths inch. In woods and damp places. Somewhat local. *H. sericea.*[71]

(*n*) Shell subglobose, depressed, thin, greenish-horn coloured, transparent, scantily covered with short hairs; whorls four and a-half, convex, the body-whorl occupying two-thirds of the shell; spire blunt, not much raised; suture deep; mouth somewhat

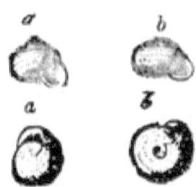

Fig. 33.—(*a*) *Helix sericea* and (*b*) *Helix hispida.*

circular, not provided with a rib; umbilicus minute. Height $\frac{1}{5}$th inch. Among nettles. Very local. *H. relevata.*

(*o*) Shell subconical, depressed, faintly keeled, horn-coloured, smooth, strongly corrugated in line of growth, thin, transparent; whorls five and a-half, convex; spire blunt, somewhat produced; suture shallow; mouth crescent-shaped; umbilicus very small. Height $\frac{1}{5}$th inch. Among decayed leaves in woods, especially those of the sycamore and alder. More or less local.

H. fusca.

(*p*) Shell subglobular, yellowish-white, marked with chocolate-brown bands and dots, giving the shell a mottled look, solid; whorls five to five and a-half, the body-whorl occupying two-thirds of the shell; spire blunt; mouth rose-coloured, crescent-shaped; umbilicus small. Height $\frac{14}{25}$ths inch. Sandhills near the sea. Local. *H. Pisana.*[72]

(*q*) Shell conical, somewhat globular, cream-coloured, with a single dark brown band on the periphery of the body-whorl, and

---

[71] v. *cornea*, horn-coloured, thin, labial rib perceptible on the outside.
[72] v. *alba*, white or pale yellow.

several bands below it; whorls six, convex; mouth rounded, purplish-brown interiorly; umbilicus narrow, deep. Height $\frac{12}{25}$ths inch. Downs and sandhills near the coast. Local.

H. virgata.[73]

(*r*) Shell subconical, depressed, whitish or dull yellowish-white, banded with brown, solid; whorls six, periphery bluntly keeled; spire slightly raised, with the apex tipped with brown; mouth roundish, with a strong white internal rib; umbilicus large. Height $\frac{7}{25}$ths inch. Under stones and among grass on sandy soils. Somewhat local. *H. caperata.*[74]

(*s*) Shell circular, depressed, greyish or whitish, banded with chestnut-brown; whorls six, spire slightly raised with a brown apex; periphery not keeled; mouth almost circular; umbilicus large and open, exposing three or four whorls. Height $\frac{8}{25}$ths inch. Heaths and downs near the sea. Local.

H. ericetorum.[75]

(*t*) Shell depressed, circular, yellowish or reddish, regularly marked with rufous brown; whorls six to seven, outer margin strongly keeled; spire slightly produced; mouth crescent-shaped; umbilicus large and deepish. Height $\frac{3}{25}$ths inch. Under stones and among decaying leaves. Common. *H. rotundata.*[76]

(*u*) Shell sub-conical, depressed below, transverse striæ oblique and strong, blackish-brown; whorls five to six, with the spire somewhat produced; mouth nearly circular, gibbous; umbilicus

---

[73] v. *subglobosa*, smaller, with a double band above the periphery, umbilicus wider; v. *subaperta*, whitish, spire more depressed, umbilicus wider; v. *leucozona*, violet or reddish-brown, whitish band at periphery; v. *submaritima*, smaller, darker, slightly pyramidal; v. *carinata*, strongly keeled, yellowish white, compressed above.

[74] v. *ornata*, smaller, whitish, with one brown band above, and two to six below; v. *major*, larger; v. *Gigaxii*, smaller, more depressed, reddish, spotted, marked by two bands below; v. *subscalaris*, conical, whorls more convex; v. *alba*, pure white, unicolor.

[75] v. *instabilis*, smaller, darker, spire more raised, umbilicus narrower, sometimes streaked or spotted; v. *alba*, white, unicolor; v. *lutescens*, dirty-yellowish, unicolor; v. *minor*, smaller.

[76] v. *alba*, white unicolor; v. *Turtonii*, almost flat above; v. *minor* smaller; v. *pyramidalis*, subconical, spire more raised.

large, deep. Height $\frac{2}{25}$ths inch. On old walls and castle-ruins. Somewhat local. *H. rupestris.*[77]

(*v*) Shell very minute, depressed, nearly circular, yellowish horn-colour; whorls four, with the spire somewhat produced; mouth semilunar; umbilicus large, deep. Height $\frac{1}{37}$th inch. Among dead leaves, and under stones. Moderately common.

H. pygmæa.

(*w*) Shell nearly circular, sub-depressed, milk-white; whorls three and a-half, the body-whorl exceeding in size the rest of the shell; spire slightly produced; mouth circular, thickened and reflected so as to form a double peristome; umbilicus large, deep. Height $\frac{3}{30}$ths inch. Among stones, moss, and grass. Somewhat local. *H. pulchella.*[78]

(*x*) Shell sharply keeled at the periphery, circular, depressed, greyish or horn-colour, streaked regularly with rufous brown; whorls five, with the spire very slightly produced; mouth oval, surrounded by a thick white reflected rim, somewhat angulated above and below; umbilicus large, deep. Height $\frac{7}{25}$ths inch. In hedgerows and woods in calcareous districts. Local.

H. lapicida.[79]

(*y*) Shell sub-discoidal, depressed, solid, dull rufous-brown hispid; whorls six and a-half; spire sunk below the level of the body-whorl; mouth triangular, marked with three protuberances, and with a pinkish-white, reflected peristome; umbilicus large, deep. Height $\frac{6}{25}$ths inch. Among the moss near the roots of shady trees at Ditcham Wood, Hants. Rare. *H. obvoluta.*

(*z*) Shell turreted, conical, yellowish-white streaked with brown, and with generally a reddish-brown or blackish band below the periphery; whorls eight to nine; spire tapering, blunt; mouth oval; umbilicus narrow, rather shallow. Height $\frac{3}{5}$ths inch. On cliffs and downs near the coast.

H. (Cochlicella) acuta (= *Bulimus acutus*).[80]

---

[77] v. *viridescenti-alba*, greenish-white.

[78] v. *costata*, not so glossy, reddish with numerous transverse, membranous ribs as well as intermediate striæ.

[79] v. *albina*, whitish, unicolor; v. *minor*, dwarfed.

[80] v. *inflata*, slightly ventricose; v. *strigata* with broad, whitish ribs

19. **Bulimus.**—Shell oval or oblong-ovate; spire blunt, but much more prominent than in *Helix;* mouth an elongated oval; umbilicus very small. Tentacles shorter than in *Helix.*

(*a*) Shell oblong, conical, light brown or dark brown; whorls six and a-half to seven and a-half, compressed, the body-whorl occupying slightly less than half the shell; spire tapering, blunt; mouth oval, pinkish or brownish inside, and with a thick peristome which is reflected over the umbilicus; umbilicus narrow, deep. Height $\frac{3}{8}$ths inch. Among fallen beech leaves in the southern and western counties. *B. montanus.*

(*b*) Shell oblong, oval, brown, more glossy than *B. montanus;* whorls six and a-half, rounded; spire elongated, blunt; mouth oval, with a white, reflected peristome; umbilicus narrow, but not deep. Length $3\frac{1}{2}$ lines. Among moss and stones in woods and hedge-banks. Moderately common. *B. obscurus.*[81]

20. **Pupa.**—Shell cylindrical or oblong, with many narrow whorls; mouth oval or lunate, generally toothed within; peristome incomplete, thickened, reflected; umbilicus very minute.

(*a*) Shell sub-cylindrical, solid, rufous-brown, conical; whorls eight to nine, spire produced; mouth semi-oblong, with eight or nine denticles; umbilicus small, oblique. Height $\frac{8}{28}$ths inch. On rocks and at roots of trees in calcareous districts. Local.

P. secale.[82]

(*b*) Shell ovate, conical, sub-cylindrical, light brown or yellowish horn-colour; whorls six to six and a-half; mouth sub-semi-circular, with seven or eight reddish plications; umbilicus oblique, distinct, narrow. Height $\frac{2}{28}$ths inch. Among moss and dead leaves, under stones. Local. *P. ringens.*

(*c*) Shell oblong, sub-cylindrical, glossy, yellowish-brown or dark horn-colour; whorls six to seven, rounded; mouth semi-oval, with a single denticle at the angle formed by the junction of the outer lip, which is broad and much reflected; umbilicus

---

alternating with brownish-grey streaks; v. *bizona*, greyish or whitish, striæ obliterated, with two dark bands on the body-whorl.

[81] v. *albinos*, whitish, unicolor.

[82] v. *edentula*, peristome without teeth; v. *alba*, white; v. *Boileausiana*, larger denticle entirely double, with a supernumerary projecting denticle at the angle of the columellar border.

small, narrow. Height $\frac{4}{25}$ths inch. Among moss and dead leaves; under stones and the bark of old trees. Common.

P. umbilicata.[83]

(*d*) Shell cylindrical, ovate, finely striated, dull brownish horn-colour; whorls six to seven, convex; mouth sub-semicircular, provided with a small denticle placed in the middle of the base of the last whorl, and a strong, white, external rib behind the outer lip; umbilicus narrow. Height $\frac{7}{80}$ths inch. In moss and under stones on sandy soils. Somewhat local.

P. marginata.[84]

21. **Vertigo.**—Shell dextral or sinistral, sub-cylindrical with closely pressed, gradually-enlarging whorls; mouth more or less angular, generally toothed internally; umbilicus minute. Resembles *Pupa*, but differs in having the ventral tentacles wanting, and in having the peristome thinner.

(*a*) Shell dextral, oval, shining, chestnut-brown; whorls four and a-half, the body-whorl comprising about one-half of the shell; mouth subcordate, with from six to eight reddish-brown denticles; umbilicus narrow, distinct. Height $\frac{1}{11}$th inch. Under stones in marshy places. Moderately common.

V. antivertigo.[85]

(*b*) Shell dextral, oval, ventricose, light yellowish horn-colour; whorls four and a-half, globose, the body-whorl comprising more than half of the shell; suture exceedingly deep; mouth semi-oval with four or five denticles; umbilicus open. Height $\frac{1}{10}$th inch. Rare.

V. Lilljeborgi.[86]

(*c*) Shell resembling *V. Lilljeborgi*, but more swollen and barrel-shaped, and with the labial rib much stouter. Rare.

V. Moulinsiana.

(*d*) Shell dextral, ovate, somewhat cylindrical, reddish-brown; whorls four and a-half to five, the body-whorl occupying about one-half of the shell, convex; mouth semi-oval with four or five

---

[83] v. *edentula*, without denticles; v. *albina*, entirely white.

[84] v. *albina*, whitish, unicolor; v. *bigranata*, with a dentiform palatal callosity.

[85] v. *sexdentata*, with six denticles; v. *octodentata*, with eight denticles.

[86] v. *bidentata*, labial or palatal teeth wanting.

teeth; umbilicus rather deep, narrow. Height $\frac{1}{10}$th inch. Under stones and logs of wood. Moderately common.
V. pygmæa. [87]

(*e*) Shell somewhat like that of *V. pygmæa* but differing in being more cylindrical, lighter in colour, more strongly striated, and in the absence of any rib. Among grass and dead leaves in the northern counties. Local. *V. alpestris.*

(*f*) Shell dextral, shortly ovate, glossy, thin, light yellowish-horn colour; whorls four and a-half, very ventricose, with the spire very short and abrupt; suture very deep; mouth sub-pyriform with six teeth; umbilicus small, contracted by a basal ridge. Height $\frac{2}{25}$ths inch. Among grass, dead leaves, and under stones. Moderately common. *V. substriata.*

(*g*) Shell dextral, ovate, cylindrical, pale brownish-horn coloured; whorls five to six and a-half; spire conical, blunt; mouth subsemi-oval, without teeth; umbilicus narrow, rather deep. Height $\frac{1}{7}$th inch. Among moss and dead leaves, under stones. Local. *V. edentula.*

(*h*) Shell dextral, cylindrical, glossy, horn-colour, narrower and smaller than *V. edentula;* whorls five and a-half, increasing in size to the third, and then becoming of the same breadth; suture deepish; mouth oblong, without teeth; umbilicus narrow, oblique. Height $\frac{2}{25}$ths inch. Under stones on hillsides. Very local.
V. minutissima.

(*i*) Shell sinistral, somewhat fusiform, glossy, thin, pale yellowish-horn colour; whorls four and a-half to five, ventricose; suture deep; mouth subquadrate, with six or seven denticles; umbilicus small, contracted by a basal ridge. Height $\frac{2}{25}$ths inch. Among moss and decayed leaves, under stones. Very local. *V. pusilla.*

(*j*) Shell sinistral, fusiform, pale horn-colour, glossy, smaller and narrower than *V. pusilla;* whorls four and a-half, the penultimate whorl being the largest; suture deepish; mouth triangularly subcordate, with from four to five denticles; umbilicus small, contracted by a basal ridge. Height $\frac{3}{50}$ths inch. At the roots of grass. Extremely local. *V. angustior.*

22. **Balia.**—Shell sinistral, elongated, thin; mouth ovate,

---

[87] v. *pallida*, thinner, lighter in colour.

sometimes with a denticle at the base of the penultimate whorl; peristome thin, umbilicus narrow.

(a) Shell fusiform, turreted, glossy, yellowish horn-colour streaked with white; whorls seven to eight; suture deepish; mouth rounded, pear-shaped, with sometimes a denticle near the middle of the base of the penultimate whorl; umbilicus small. Height $\frac{9}{25}$ths inch. Under the bark of old trees. Local.

B. perversa.[88]

23. **Clausilia.**—Shell sinistral, fusiform; mouth pyriform or elliptical, toothed, furnished with a clausilium; umbilicus very small.

(a) Shell cylindrical, glossy, chocolate-brown or horn-coloured, marked with whitish streaks; whorls ten to thirteen; mouth pyriform, furnished with two plications on the base of the penultimate whorl with one to three smaller plaits between them, and with two plications on the columella; umbilicus narrow; clausilium oblong. Height $\frac{3}{5}$ths inch. Under stones, logs of wood, and among grass on hedge-banks, on trunks of beech, willow, and ash trees. Common. *C. rugosa.*[89]

(b) Shell fusiform, reddish or yellowish-brown; whorls nine to ten, swollen; mouth rounded, pyriform, with plaits as in *C. rugosa*, but the lower one on the base of the penultimate whorl is less prominent and sometimes cruciate; umbilicus indistinct; clausilium oblong. Length $\frac{13}{25}$ths inch. Among dead leaves and under the bark of tress. Rare. *C. Rolphii.*

(c) Shell fusiform, yellowish-brown, streaked with white; whorls twelve to thirteen, compressed; mouth pear-shaped, canaliculated, with the same plaits as in the two foregoing species, except that the teeth between the plications on the base of the penultimate whorl are not represented; umbilicus rather broad, basal crest prominent, clausilium oval. Length $\frac{17}{25}$ths inch. On the bark of willow trees. Very local. *C. biplicata.*

[81] v. *viridula*, greenish-white, transparent.

[89] v. *dubia*, larger, more ventricose; v. *gracilior*, longer, more slender; v. *tumidula*, smaller, shorter, more ventricose; v. *Everetti*, shorter, whorls fewer; v. *Schlechtii*, like v. *dubia* but larger, elongated, smoother, and more transparent, pale brown. (Possibly v. *dubia* is a distinct and separate species.)

(*d*) Shell fusiform, large, yellowish-brown, glossy; whorls twelve; mouth oval, pear-shaped, with two very strong plaits on the base of the penultimate whorl, and two or three labial plaits which are distinctly visible through the shell; umbilicus small; basal crest slight; clausilium with a notch on the margin near its

Fig. 34.—*Clausilia biplicata*.

base. Height $1\frac{7}{25}$ths inch. On beech and ash trees. Moderately distributed, but local. *C. laminata*.[90]

23. **Cochlicopa**.—Shell oblong or oblong oval, very glossy, transparent; mouth pear-shaped; outer lip thickened, not reflected.

(*a*) Shell fusiform, pale yellowish-brown, glossy; whorls seven; spire produced, blunt; suture shallow; mouth pear-shaped, with three or more plications. Length $\frac{2}{25}$ths inch. Among moss and decaying leaves. Moderately common, but local. *C. tridens*.[91]

(*b*) Shell oblong, not so ventricose as *C. tridens*, yellowish-brown or greenish-white, glossy; whorls five to five and a-half; suture rather deep; mouth oval, without plaits. Height $\frac{6}{25}$ths inch. Among moss and decaying leaves. Common. *C. lubrica*.[92]

[90] v. *albinos*, entirely white; v. *pellucida*, thinner, more transparent, very glossy.

[91] v. *crystallina*, greenish-white, transparent, glossy.

[92] v. *hyalina*, greenish-white; v. *fusca*, more or less a deep brown colour; v. *lubricoides*, smaller, more slender; v. *viridula*, shaped like v. *lubricoides*, but greenish-white; v. *ovata*, smaller, oval, spire shorter.

25. **Achatina.**—Shell cylindrical, smooth, thin, glossy; mouth oval, with a notch at its base; outer lip thin and not reflected; umbilicus absent.

(*a*) Shell fusiform, white; whorls five and a-half to six; spire attenuated; suture deep; mouth lanceolate, sharply angled above, notched deeply at its base. Length $\frac{2}{11}$ths inch. Subterranean; specimens are often found in river ejectamenta after floods. Local. *A. acicula.*

Family V.—**Carychiidæ.**

Shell spiral, oblong; mouth oval, denticulated; umbilicus very small. Eyes situated at the posterior base of the dorsal tentacles; lower tentacles rudimentary.

26. **Carychium.**

(*a*) Shell white, fusiform, oblong; whorls five to five and a-half, the body-whorl comprising nearly one-half of the shell; apex acute; mouth ear-shaped with three denticles; umbilical chink oblique. Height $\frac{1}{5}$th inch. Under stones, in decaying leaves, and at the roots of grass. Moderately common. *C. minimum.*

Family VI.—**Cyclostomatidæ.**

Shell cylindrical or conical, operculated; mouth round or oval; operculum shelly or horny.

27. **Cyclostoma.**—Shell conical; operculum shelly.

(*a*) Shell greyish or yellowish-brown streaked and blotched with purplish-brown, coarsely striated; whorls four and a-half, tumid; suture deep; mouth subrotund; umbilicus narrow. Length $\frac{14}{23}$ths inch. In hedgerows and on hillsides in calcareous districts. Moderately common. *C. elegans.*[93]

28. **Acme.**—Shell cylindrical; operculum horny.

---

[92] v. *ochroleuca*, yellowish, unicolor, bandless; v. *pallida*, very pale yellowish, with some spots and demi-effaced bands; v. *albescens*, whitish, unicolor.

(*a*) Shell light brown, attenuated, glossy, strongly striated in line of growth; whorls six to seven; suture well defined; mouth pyriform, contracted above; umbilicus minute. Height $\frac{3}{25}$ths inch. Under stones and among decaying leaves. Moderately common, but local. *A. lineata.*[94]

[94] v. *alba*, white or colourless, transparent; m. *sinistrorsum*, spire sinistral.

CHAPTER V.

CENSUS OF THE AUTHENTICATED DISTRIBUTION OF BRITISH LAND AND FRESH-WATER MOLLUSCA.

By JOHN W. TAYLOR, F.L.S., AND W. DENISON ROEBUCK, F.L.S.

With the view of facilitating the labours of the numerous conchologists and others who are throughout the United Kingdom assisting us in working out the detailed distribution of the British land and fresh-water mollusca, and also of those who may hereafter be disposed to assist us by submitting specimens for authentication and record, the following "census" has been framed. It is strictly limited to "authenticated" records, that is, to such as are verified by specimens which have passed under the examination of the Conchological Society's referees. This process secures uniformity of value in the results.

The numbers are those of counties and vice-counties, a key to which will be found at the end of the chapter. It will be seen that the Channel Islands are denoted by the cypher "o" and that they precede district No. 1. The total number of counties and vice-counties for which we have seen specimens is given under each species.

For the present the distribution of the varieties cannot be given, and some of the doubtful species—such, for example, as the mythical *Arion flavus*—are also omitted.

Conchologists—and naturalists generally—are requested to aid in filling up the gaps in the records by sending us specimens for authentication and study. They will not fail to note how very imperfectly the mollusca of Scotland and Ireland are recorded.

Testacella maugei Fér.   Nine counties.
> Eng. S. 0 1 6 9 22 34.   Wales 41 45.   Irel. 145.   No records for Eng. N., or Scotland.

T. haliotidea F.-Big.   Eight counties.
> Eng. S. 3 13 14 22-23 25 34.   Eng. N. 61.   Irel 146.   No records for Wales or Scotland.

T. scutulum Sow.   Thirteen counties.
> Eng. S. 10 13 15—17 20-21 27-28 37.   Eng. N. 55-56 58 61 66.   Scot. 86.   No records for Wales or Ireland.

Arion ater L.   Ninety-one counties.
> Eng. S. 1—13 16—21 24—26 28—30 32—40. Wales, 41-42 45—52.   Eng. N. 54—67 69—71.   Scot. 75 79 81 83 86-87 90 92 96 98—101 103 107—109. Irel. 114—116 118 124 136 138 145-146.

Arion subfuscus Drap.   Forty-four counties.
> Eng. S. 1 2 5 9—13 21 34 39-40.   Wales 41 45 47 49.   Eng. N. 54 56—59 62 64—67 69.   Scot. 77 79 81 83 92 98 100-101 105 107.   Irel. 114 116 124 126 136 138 145.

A. hortensis Fér.   Seventy-five counties.
> Eng. S. 1 4—6 8 10—13 15—17 20—22 24 30 32—40.   Wales 41 45—52.   Eng. N. 53—59 61—66 69. Scot. 75—77 79 81—86 88 90 95 98—100 105 107 109-110.   Irel. 114 124 127 145 148.

A. bourguignati Mab.   Sixty-eight counties.
> Eng. S. 1-2 4 9 11—13 16-17 21 25 28 30 34—40. Wales 41 47—49.   Eng. N. 53—58 60—67 69.   Scot. 75—85 87-88 90 92 96 98—100 103 107-108 110.   Irel. 114 124 127 145 147-148.

Geomalacus maculosus Allm.   One county.
> Is found only in county Kerry.

Amalia gagates Dp.   Twenty-four counties.
> Eng. S. 1 3 5 9—11 14-15 21 35-36 39-40.   Wales 46.   Eng. N. 54 56 62 66 71.   Scot. 83 100.   Irel. 114 124 138.

Amalia marginata Müll.   Thirty-six counties.
> Eng. S. 1 3—6 9—13 15—17 20-21 25 33—36.

Wales 41 45-46 50. Eng. N. 53—55 62 64 71. Scot. 76 85 100. Irel. 114 124 145.

Limax agrestis L. One hundred counties.

Eng. S. 1 2 4—6 8—13 15—21 24-25 28—40. Wales 41-42 45—52. Eng. N. 53—67 69—71. Scot. 72 75—85 87-88 90 92 95-96 98—101 103 105—110 112. Irel. 114 116 119 124 126-127 136 138 145.

Limax lævis Müll. Twenty-two counties.

Eng. S. 1 11 13 16 21 39. Wales 51. Eng. N. 56 58 62—66. Scot. 76 95 98 100-101 107—109. No records for Ireland.

Limax flavus L. Thirty-four counties.

Eng. S. 1 5-6 11—13 15—17 20-21 25 29-30 33 37 —40. Wales 45 47. Eng. N. 54—56 61—64 66. Scot. 76. Irel. 124 127 145 147.

L. cinereo-niger Wolf. Thirteen counties.

Eng. S. 13 39-40. Wales 48. Eng. N. 62 64. Scot. 90 96 107. Irel. 118 126 130 145 148.

L. maximus Auct. Seventy counties.

Eng. S. 1 4-5 8-9 11—13 15—17 19—21 24-25 28 30 32—34 36—40. Wales 45—50 52. Eng. N. 53—59 61—66 69. Scot. 75-76 78—83 85 88 90 98—100 107. Irel. 114 116 124 132 136 138 145 148.

Limax arborum Bouch. Forty counties.

Eng. S. 1 2 9 11—13 18 20 23 27 33-34 36—39. Wales 45 48—50. Eng. N. 54 56—58 62—66. Scot. 81 88 107 109. Irel. 115 118 136 138 142 145 148.

Succinea putris L. Fifty-six counties

Eng. S. 0-1 6 11—17 19—21 23—25 27 29 31—35 37—40. Wales 41 44-45 49—52. Eng. N 53—59 61—67 69. Scot. 73 76 78 89-90. Irel. 144 148.

S. elegans Risso. Twenty-three counties.

Eng. S. 0 6 10-11 13 15—17 20-21 23 32—34. Wales 44-55 50. Eng. N. 56 58 61—64. No records from Scotland or Ireland.

S. virescens Morel. Three counties.

Eng. S. 16 21 37. No records from Eng. N., Wales, Scotland, or Ireland.

S. oblonga Drap.   Four counties.
   Wales 41.   Eng. N. 62.   Scotland 84.   Ireland 147.
No records for England S.

Vitrina pellucida Müll.   Sixty-eight counties.
   Eng. S. 1 3-4 6 11 13—17 20 22 23 25 27-28 32—34 36—40.   Wales 41 45 47 49 52.   Eng. N. 54—59 61—67 69 71.   Scot. 75—83 85 89 90—92 98—101 107 109. Irel. 114 132 138 144.

Zonites cellarius Müll.   Eighty-eight counties.
   Eng. S. 0—9 11—13 16—18 20—22 25 27—30 32—40.   Wales 41 43 45 47—52.   Eng. N. 53—59 61—67 69 71.   Scot. 73 75—85 89—92 98—101 105 107 109.   Irel. 114 118 124 132 138 144-145.

Z. alliarius Miller.   Forty-four counties.
   Eng. S. 0-1 6 11 13 16-17 20 22 32—36 38 40. Wales 45 47 49—52.   Eng. N. 54 56—58 61—67 69 71.   Scot. 85 88—90 107 109 112.   Irel. 113 144.

Z. glaber Stud.   Twenty-four counties.
   Eng. S. 1 16-17 20—22 30 32 40.   Wales 44 47 49-50 52.   Eng. N. 55—60 63—66.   No records for Scotland or Ireland.

Z. nitidulus Drap.   Sixty-eight counties.
   Eng. S. 0-1 4 6 8—13 15—18 20—24 29 32—40. Wales 41 43—45 47—52.   Eng. N. 53—67 69.   Scot. 73 76 83 85 89-90 98 102 104 109.   Ireland. 113 138 144.

Z. purus Ald.   Thirty-seven counties.
   Eng. S. 1 8 11 16-17 19-20 22 25 29 32 34—37 39. Wales 49 50 52.   Eng. N. 56—59 62—67 69.   Scot. 73 89-90 107 109.   Irel. 113 144.

Z. radiatulus Ald.   Twenty-one counties.
   Eng. S. 6 13 17 21 32 34-35.   Wales 48 52.   Eng. N. 56—58 62—67 70.   Scot. 89 109.   No records for Ireland.

Z. nitidus Müll.   Twenty-four counties.
   Eng. S. 6 11 13 16-17 20—22 25 34-35 39-40.   Wales 41 51-52   Eng. N. 54 56—59 62—64.   No records for Scotland or Ireland.

Z. excavatus Müll. Thirteen counties.
 Eng. S. 1 3 11 13 23. Wales 45 48. Eng. N. 57 63-64 66. Scot. 100. Irel. 124.

Z. crystallinus Müll. Forty-five counties.
 Eng. S. 3 6 11—13 16-17 19—22 25 32—34 36 39. Wales 47 49—52. Eng. N. 54—67 69. Scot. 73 76 90 107 109. Irel. 113 144 148.

Z. fulvus Müll. Fifty counties.
 Eng. S. 7 11—13 17 20 22 25 27 32-33 36—40. Wales 41 45 47—52. Eng. N. 54 56—59 61—66 69-70. Scot. 76 83 88—90 98 100-101 107 109. Irel. 113-114 118 148.

Helix lamellata Jeff. Eleven counties.
 Eng N. 57 62—64 66-67. Scot. 72 88 98 107. Irel. 148. No records for Wales or England South.

H. aculeata Müll. Thirty-three counties.
 Eng. S. 9 11 14 16-17 20 22-23 32-33 40. Wales 48-49 52. Eng. N. 56— 60 62—66 69. Scot. 75-76 89-90 107 109. Irel. 113 148.

H. pomatia L. Eight counties.
 Eng. S. 12-13 15—17 23 30 33. No records for Wales, England N., Scotland, or Ireland.

H. aspersa Müll. Seventy-seven counties.
 Eng. 0—6 8—21 23—25 27—37 39-40. Wales 41-42 44—52. Eng. N. 53—58 60—71. Scot. 73 82-83 85 89-90 94. Irel. 114 119 124 144.

H. nemoralis L. One hundred and three counties.
 Eng. S. 0—6 8—25 27—40. Wales 41-42 44—52. Eng. N. 53—69 71. Scot. 73 75-76 78—83 85-86 88 90 92 96 98—103. Irel. 114-115 118-119 124-125 131-132 136 138-139 141 144 148.

H. hortensis Müll. Seventy-nine counties.
 Eng. S. 1—6 8—23 25 28—40. Wales 41—45 47-48 50. Eng. N. 54—59 61—67 69. Scot. 73 75—77 81-82 85 88—90 97 99 102-103 107—109. Irel. 118 124-125 132.

H. arbustorum L. Fifty counties.
 Eng. S. 2 6 10-11 13—17 20—23 25 31—37 39. Wales 41 43—45 49-50. Eng. N. 53-54 56—57 59 61—67 69. Scot. 72 77 88 90 98 107—109 112. No records for Ireland.

CENSUS. 99

H. **cantiana** Mont. Thirty-seven counties.
   Eng. S. o 5-6 10—20 22-23 25 27—30 32 34 36—38.
   Wales 41. Eng. N. 54—56 61—67. No records for Scotland or Ireland.

H. **carthusiana** Müll. Three counties.
   Eng. S. 13—15.

H. **rufescens** Penn. Sixty-three counties.
   Eng. S. 1—17 19—24 26 28—30 32—35 37 39-40.
   Wales 41 44-45 48—52. Eng. N. 54 57—65 69-70
   Scot. 76 98-99. Irel. 113 119 124 144 146 148.

H. **concinna** Jeff. Thirty-three counties.
   Eng. S. o 6 8 11 14—17 20—24 31—34 38-39.
   Wales 45 49-50 52. Eng. N. 53 56-57 62 64-65 67.
   Scot. 73 77 89. No records for Ireland.

H. **hispida** L. Fifty-eight counties.
   Eng. S. 0-1 6 8-9 11—18 20-21 23-24 29—36 39-40.
   Wales 41-42 47—52. Eng. N. 53—67 69 71. Scot. 73
   76 88—90. Irel. 141.

H. **sericea** Müll. Thirty counties.
   Eng. S. 1—3 11-12 15 20-21 29 32—34. Wales 44-
   45 49—52. Eng. N. 56 62 64—67 69. Scot. 76 90 98
   101. Irel. 125.

H. **revelata** Mich. Four counties.
   Eng. S. o—3. No records for Scotland, Wales,
   Ireland, or England N.

H. **fusca** Mont. Twenty-four counties.
   Eng. S. 3 11 13 34 36 39. Wales 47—49. Eng. N.
   57 59 62—64 66-67. Scot. 73 75 80 107. Irel. 114 118
   138 145.

H. **pisana** Müll. Five counties.
   Eng. S. 0-1. Wales 45. Ire. 123-124. No records
   for Scotland or North of England.

H. **virgata** DaCosta. Fifty-eight counties.
   Eng. S. o—11 13—23 25 27—29 31—34 37-38.
   Wales 41 44-45 49—52. Eng. N. 53—57 60—64 66.
   Scot. 75. Irel. 114 124-125 139 144 148.

H. **caperata** Mont. Sixty-four counties.
   Eng. S. 0-1 3-4 6 9—23 25 32—34 36—40. Wales

41 44-45 48—52. Eng. N. 53—59 61—67 69 71. Scot. 75-76 83 85 90 100 107. Irel. 113 124 144.

H. ericetorum Müll. Sixty counties.

Eng. S. 1 6—8 11—17 19-20 22-23 27—30 32—34 37—40. Wales 41 44-45 50 52. Eng. N. 53—57 61—67 71. Scot. 85 97 100 103 107-108 110. Irel. 113-114 119 124 127 131 139 141 144.

H. rotundata Müll. Eighty-six counties.

Eng. S. 0-1 4 6—11 13—17 20—23 25 27-28 30 32—37 39-40. Wales 41 43 45 47—52. Eng. N. 53—59 61—67 69—71. Scot. 73 75—79 81—85 88—91 97—101 105 107 109. Irel. 113-114 118 124 132 136 144.

H. rupestris Dp. Thirty-six counties.

Eng. S. 0-1 3 6 9 11 13—15 20 22-23 32—34 37 39. Wales 41 49-50 52. Eng. N. 57 60 62 64—66 69—71. Scot. 89. Irel. 124 132 144 146 148.

H. pygmæa Dp. Twenty-two counties.

Eng. S. 11 15 23 32-33 39. Wales 50. Eng. N. 54 56 58 61 64—67 71. Scot. 89 105 107—109. Irel. 113.

H. pulchella Müll. Forty-five counties.

Eng. S. 0-1 6 11 13—16 19 22-23 25 29 32—36 39. Wales 41 49—52. Eng. N. 53—57 59 61—66 69. Scot. 83 85 89—91 107. Irel. 139 144.

H. lapicida L. Thirty-four counties.

Eng. S. 4 6 9 11—13 15—19 22—24 27 30 32—37 39. Wales 41 43. Eng. N. 53 56-57 59 61—65. No records for Scotland or Ireland.

H. obvoluta Mull. Three counties.

Eng. S. 11—13. No records for Wales, Ireland, Scotland, or North of England.

Bulimus acutus Müll. Twenty-five counties.

Eng. S. 0-1 3-4 6 9-10. Wales 41 44-45 50—52. Eng. N. 58 71. Scot. 103. Irel. 113-114 119 124 127 132 134 139 144.

B. montanus Dp. Five counties.

Eng. S. 6 23 26 33-34. No records for England N., Wales, Scotland, or Ireland.

CENSUS. 101

B. obscurus Müll. Fifty-four counties.

Eng. S. 0-1 3-4 6 9 11—17 19-20 22-23 27—30 32—36 38—40. Wales 41 50 52. Eng. N. 53-54 56—67 69. Scotland 77 81 83 88—91. No records for Ireland.

Pupa secale Dp. Sixteen counties.

Eng. S. 6 9-10 12—15 22-23 33—35. Eng. N. 57 64-65 69. No records for Wales, Scotland, or Ireland

P. ringens Jeff. Fourteen counties.

Eng. S. 0 36. Wales 52. Eng. N. 56 61-62 64 66-67. Scot. 76 101 107-108. Irel. 148.

P. umbilicata Dp. Sixty-four counties.

Eng. S. 0—3 6 9 11—17 21—23 25 32—34 37—40. Wales 41 44-45 47 49—52. Eng. N. 54 56—67 69—71 Scot. 73 75 83 85 89-90 98 107—109. Irel. 119 139 144 146—148.

P. marginata Drap. Thirty-seven counties.

Eng. S. 1 4 6 9—17 19 21 23 25 32—34 37. Wales 49—51. Eng. N. 54 56—58 61—64 66-67 71. Irel. 139 144 148. No records for Scotland.

Vertigo antivertigo Dp. Nineteen counties.

Eng. S. 4 10-11 17 22-23 25 33-34. Wales 50. Eng. N. 56-57 59 62—64. Scotland 81 107. Ireland 124.

V. lilljeborgi Westerl. One county.

Irel. 139. No records for Eng. Wales, or Scotland.

V. moulinsiana Dup. non Jeff. One county.

Eng. S. 20. No records for England N., Wales, Scotland, or Ireland.

V. pygmæa Drap. Thirty-four counties.

Eng. S. 4 6 10—15 17 21 23 25 27 30 32—34 38. Wales 49. Eng. N. 54 56-57 62—67. Scot. 76 83 107 109. Irel. 113 148.

V. alpestris Ald. Four counties.

Eng. N. 63 67 69. Irel. 113. No records for England S., Wales, or Scotland.

V. substriata Jeff. Five counties.

Wales 52. Eng. N. 57 63 66. Irel. 113. No records for England S., or Scotland.

V. pusilla Müll. Ten counties.
  Eng. S. 4 23. Eng. N. 56-57 63-64 66-67. Scot. 75. Irel. 113. No records for Wales.
V tumida Westerl. Not seen.
V. angustior Jeff. Six counties.
  Eng. N. 57 64. Scot. 107. Irel. 113 136 139 147. No records for England S. or Wales.
V. edentula Drap. Thirty-five counties.
  Eng. S. 4 11 20 23 28 32—34 37 39. Wales 48 52. Eng. N. 54 56-57 60 62—64 66-67 69. Scot. 76 81 89—91 100-101 107—109. Irel. 113-114 148.
V. minutissima Hartm. Five counties.
  Eng. S. 10. Eng. N. 63 66. Scot. 83. Irel. 148. No records for Wales.
Balea perversa L. Forty-one counties.
  Eng. S. 0 3-4 6-7 10—14 22 25 27 32-33 38 40. Wales 41 45 49-50. Eng. N. 54-55 57—60 62 64-65 69—71. Scot. 73 75-76 83 89 98 101 109. Irel. 148.
Clausilia rugosa Drap. Eighty-nine counties.
  Eng. S. 0—4 6—9 11—25 27—30 32—40. Wales 41—45 47—52. Eng. N. 53—67 69—71. Scot. 73 76 83 88—90 98 100-101 103 107—109. Irel. 113-114 124 132 138-139 144 146—148.
C. rolphii Gray. Nine counties.
  Eng. S. 12—17 22 33. Eng. N. 56. No records for Wales, Scotland, or Ireland.
C. biplicata Mont. Three counties.
  Eng. S. 17 20-21. No records for Wales, England N., Scotland, or Ireland.
C. laminata Mont. Thirty-five counties.
  Eng. S. 6 9 11—17 20 22-23 25—27 32—34 36 39. Wales 41. Eng. N. 53-54 56—58 61—67 69. Scot. 89. No records for Ireland.
Cochlicopa tridens Pult. Twenty-two counties.
  Eng. S. 1 6 14 17 23 30 33-34 37—39. Wales 48. Eng. N. 56—58 62—67 69. No records for Scotland or Ireland.
C. lubrica Müll. Eighty-five counties.
  Eng. S. 0—2 4 6—8 10—23 25 27—30 32—40.

Wales 41 43 45 47—52. Eng. N. 54—67 69 71.
Scot. 73 75—77 79—83 85 88—90 98 100 105 107—109.
Irel. 113-114 132 139 144 148.

Achatina acicula Müll. Eighteen counties.

Eng. S. 6 8—10 14-15 17 19 23 32 34 36. Eng. N. 56-57 62—64 67. No records for Wales, Scotland, or Ireland.

Carychium minimum Müll. Fifty counties.

Eng. S. 6 11 13—17 19—23 25 27-28 32—34 36 39. Wales 41 48—52. Eng. N. 54 56—59 61—67 69. Scot. 73 76 89—91 98 101 107. Irel. 113 144 148.

Cyclostoma elegans Müll. Twenty-five counties.

Eng. S. 3-4 6 9 11—17 20 22-23 29-30 33 35-36. Wales 41 45 50. Eng. N. 62 64 69. No records for Scotland or Ireland.

Acme lineata Drap. Eleven counties.

Eng. S. 6 15 23. Eng. N. 57 60 62—64 66. Scot. 76. Irel. 148.

Sphærium corneum L. Sixty counties.

Eng. S. 6-7 9 11—25 27—30 32—34 37—40. Wales 41-42 45 47 52. Eng. N. 53—67 69. Scot. 73 76 81 83 88—90 112. Irel. 115 144.

S. rivicola Leach. Nineteen counties.

Eng. S. 6 17 21—23 32—34 37 39. Eng. N. 53 56—59 62—64 67. No records for Wales, Scotland, or Ireland.

S. ovale Fér. Seven counties.

Eng. S. 17 21 38. Eng. N. 59 62—64. No records for Wales, Scotland, or Ireland.

S. lacustre Müll. Thirty-three counties.

Eng. S. 3 6-7 11—13 15—17 19 21 27-28 32 34 37—40. Wales 41 44 49-50. Eng. N. 55—57 59 61—64 66-67. No records for Scotland or Ireland.

Pisidium amnicum Müll. Twenty-nine counties.

Eng. S. 6 11 13 17—24 27—29 38-39. Wales. 42. Eng. N. 53 55—65 67. No records for Scotland or Ireland.

P. fontinale Drap. Forty-two counties.
  Eng. S. 0-1 6-7 9—13 17 21 28 30 32—34 37-38 40. Wales 41 44-45 47. Eng. N. 53 56—67 69. Scot. 3 0 101 108. Irel. 124.

P. pusillum Gmel. Fifty-two counties.
  Eng. S. 0-1 4 6-7 9 11—17 21 23 26 28 30 32-33 36-37 39-40. Wales 41-42 45 47. Eng. N. 53—59 61—64 66-67 69-70. Scot. 73 78 81 90 101 105 107-108. Irel. 144.

P. nitidum Jen. Twenty-four counties.
  Eng. S. 6 10-11 16 23 28 37-38. Wales 41-42. Eng. N. 54 56—59 61 63—65 67. Scot. 85 94 112. Irel. 113.

P. roseum Scholtz. Thirteen counties.
  Eng. S. 0 11 15 17 20-21 30 40. Eng. N. 56-57 63 66. Scot. 85. No records for Wales or Ireland.

Unio tumidus Phil. Twenty-one counties.
  Eng. S. 6 21—23 30 32 34 37—39. Wales 47. Eng. N. 53—57 59 61—64. No records for Scotland or Ireland.

U. pictorum L. Twenty-four counties.
  Eng. S. 6 8 17 21-22 27 29-30 32 37—39. Wales 47. Eng. N. 53—59 62—64 68. No records for Scotland or Ireland.

U. margaritifer L. Sixteen counties.
  Wales 45 49-50. Eng. N. 62 67 71. Scot. 76-77 87-88 90 98 108. Irel. 126 144-145. No records for England S.

Anodonta cygnea L. Forty-one counties.
  Eng. S. 8 11 17-18 20 25 27-28 30 32—34 37—40. Wales 42 45. Eng. N. 53—67. Scot. 76 90. Irel. 115.

A. anatina L. Twenty-five counties.
  Eng. S. 6 15-16 20—22 27 32-34 38—40. Wales 45. Eng. N. 55—59 61—65 67. Scot. 83. No records for Ireland.

Dreissena polymorpha Pall. Twenty-three counties.
  Eng. S. 6 17 21—23 29 31—34 37—40. Eng. N.

55—59 63-64. Scot. 76-77. No records for Wales or Ireland.

Neritina fluviatilis L. Thirty-two counties.
Eng. S. 3 6 9 11 13-14 16 18-19 21—24 27 29 32 37—39. Eng. N. 53 55—57 60—65. Scot. 76 111. Irel. 131. No records for Wales.

Paludina contecta Millet. Nineteen counties.
Eng. S. 11 22—24 27—29 31-32. Eng. N. 53—56 58-59 61—64. No records for Wales, Scotland, or Ireland.

P. vivipara L. Twenty-eight counties.
Eng. S. 16—18 20—23 25 27 31—35 37—40. Eng. N. 53 55—59 61—64. No records for Wales, Scotland, or Ireland.

Bythinia tentaculata L. Fifty-four counties.
Eng. S. 1 4—7 9—12 15—25 27—30 32—34 37—40. Wales 41-42 47. Eng. N. 53—67. Scot. 76. Irel. 113 115 142 144.

B. leachii Shepp. Twenty-eight counties.
Eng. S. 6 11 14—16 19 21 23 25 27—29 32—34 39-40. Wales 45. Eng. N. 53-54 56-57 59—64. No records for Scotland or Ireland.

Valvata piscinalis Müll. Fifty counties.
Eng. S. 1 6 9 11-12 15—25 27-28 32—34 38—40. Wales 41 45 52. Eng. N. 53—57 59—67 71. Scot. 73 76 83 85 89-90. Irel. 142—144.

V. cristata Müll. Thirty-two counties.
Eng. S. 6 11—19 21—23 25 27-28 32. Wales 41-42 45. Eng. N. 54 56-57 59 61—64 66-67. Scot. 73 77 90. No records for Ireland.

Planorbis lineatus Walk. Eleven counties.
Eng. S. 14—17 21—23 27-28. Eng. N. 60-61 64. No records for Wales, Scotland, or Ireland.

P. nitidus Müll. Thirty counties.
Eng. S. 6-7 11 13 16-17 19 21 27-28 31 37—40. Wales 41. Eng. N. 53 55-56 58-59 61—64 67. Scot. 76 85. Irel. 142 147.

P. nautileus L. Thirty-six counties.
Eng. S. 3 6 9 11 16 19 21 23 27-28 30 32 37—39.

Wales 41. Eng. N. 53-54 56—65 67. Scot. 73 76-77 82 85 90 100. No records for Ireland.

P. albus Müll. Fifty-three counties.
    Eng. S. 3 6-7 10-11 13 15—19 21—23 25 27-28 30 32—35 37—40. Wales 41 45 47 52. Eng. N. 53—65 67 71. Scot. 73 76 82 85 90 100. Irel. 144 147.

P. parvus Say. Twenty-three counties.
    Eng. S 3 6 9 13 17-18 21 23 38. Eng. N. 55—57 59 63 65—68. Scot. 82 90 108. Irel. 114 144. No records for Wales.

P. dilatatus Gld. One county.
    Eng. N. 59. No records for Wales, Scotland, or Ireland.

P. spirorbis Müll. Fifty-two counties.
    Eng. S. 0-1 6-7 10—13 16 18—23 27—29 32-33 37—40. Wales 41-42 45 48 52. Eng. N. 53—64 66-67. Scot. 73 76 89 90 92 103 108. Irel. 142 144.

P. vortex L. Forty-five counties.
    Eng. S. 0 3 6-7 9 11 13 15—25 27—29 32—34 37—40. Wales 41 42 47. Eng. N. 53—59 61—64 Scot. 90. Irel. 131 142.

P. carinatus Müll. Forty-three counties.
    Eng. S. 6 9—11 15 17—25 27—29 32—34 37—39. Wales 41-42. Eng. N. 53—59 61—66. Scot. 83 Irel. 113 115 131 142.

P. complanatus L. Forty-eight counties.
    Eng. S. 5—7 11 13—17 19—30 32—34 37—40. Wales 41-42. Eng. N. 53—57 59—67. Scot. 83. Irel. 131 142 144.

P. corneus L. Thirty-five counties.
    Eng. S. 6 11 13 15—17 20—24 27—34 37—39. Eng. N. 53—59 61—65. Irel. 142. No records for Wales or Scotland.

P. contortus L. Forty-four counties.
    Eng. S. 6-7 10-11 13 15—17 19—23 27—30 32-33 37—39. Wales 52. Eng. N. 53—64 66. Scot. 73 85 89-90 92 100. Irel. 142 144.

Physa hypnorum L. Thirty-two counties.
    Eng. S. 0 6 11 16 19 29 34 37—40. Wales 41-42

45 50. Eng. N. 55—67. Scot. 81 103. Irel. 146 148.

P. fontinalis L. Fifty-three counties.
    Eng. S. 6 9 11 13 15—17 19—23 25 27—30 32-33 37—40. Wales, 41-42 45 52. Eng. N. 53—59 61—64 66-67 69 71. Scot. 73 76-77 83 85 89-90 100. Irel. 142 144 148.

Limnæa glutinosa Müll. Eight counties.
    Eng. S. 15 22 24 27. Eng. N. 62 69. Irel. 115 131. No records for Wales or Scotland.

L. involuta Thomps. One county.
    Confined to county Kerry (No. 148).

L. peregra Müll. Ninety-six counties.
    Eng. S. 0-1 4—8 10—34 36—40. Wales 41-42 44-45 48—50 52. Eng. N. 53—71. Scotland 72-73 75—83 85 88—90 92 98 100-101 103 105 107-108 110 112. Irel 113—115 131 141-142 144 148.

L. auricularia L. Forty-six counties.
    Eng. S. 6-7 11 13 15—18 21—25 27-28 30 32—34 37—40. Wales 42 44 52. Eng. N. 53—59 61—64 66-67. Scot. 73 76 81—83 85. Irel. 142.

L. stagnalis L. Forty-eight counties.
    Eng. S. 5-6 8-9 11 13 15—17 19—25 27—30 32—34 37—40. Eng. N. 53—59 61—67. Scot. 77. Irel. 115 136 142. No records for Wales.

L. palustris Müll. Seventy counties.
    Eng. S. 0 1 6-7 9 11—23 25 27—29 31—34 37—40. Wales 41-42 45 47 52. Eng. N. 53—67 69—71. Scot. 73 76-77 85 88—90 93. Irel. 113 115 124 131 136 142 144-145 148.

L. truncatula Müll. Seventy-four counties.
    Eng. S. 1 3-4 6 9—11 13 15—19 21—23 25 27-28 30—34 36—40. Wales 41 44 47—50 52. Eng. N. 53—67 69—71. Scot. 73 75—77 85 89—91 100-101 103-104 107—109. Irel. 113 139 141-142 144.

L. glabra Müll. Twenty-two counties.
    Eng. S. 0 11 16 19 37 39-40. Wales 47-48. Eng. N. 53 59 61—67 69-70. Scot. 77. Ireland 147.

Ancylus fluviatilis Müll. Sixty-seven counties.
Eng. S 1-2 4 6 8-9 11 13—17 19-20 23 28 32 34—40.
Wales 41 44-45 47—50 52. Eng. N. 54—59 61—65 67 69—71. Scot. 73 75-76 78-79 89-90 92 98—101 104 108. Irel. 113 115 124 136 142 144.

A. lacustris L. Forty-one counties.
Eng. S. 6-7 11 16 19 21—23 27—34 37—40. Wales N. 41 45. Eng. 54—67. Scot. 73 85 90. Irel. 124 142.

It will be seen on perusal of the above census that the principal places to be regarded as among the uninvestigated and "dark corners" of the kingdom are Scotland and Ireland, Mid-Wales, and the eastern and south-western counties of England, specimens from any of which districts will be of unusual value and importance.

KEY TO THE ABOVE.

0, Channel Isles; 1, Cornwall West, divided from Cornwall East by the highroad from Truro through St. Columb to the inland extremity of Padstow Creek; 2, Cornwall East; 3, Devon South, divided from North Devon by the watershed line, which commences at the Tamar, about midway between Tavistock and Launceston, passes over the ridge of Dartmoor, and joins the Western Canal at Tiverton; 4, Devon North; 5, Somerset South, divided from North Somerset by the River Parret from Bridgewater to Ilchester, the line thence curving round to the N. extremity of Dorsetshire; 6, Somerset North; 7, Wilts. North, separated from Wilts. South by the Kennett and Avon Canal; 8, Wilts. South; 9, Dorsetshire; 10, Isle of Wight; 11, Hants South, divided from Hants North by the highroads running W. and E. to the borders of Wilts and Sussex respectively, through Stockbridge and Petersfield; 12, Hants North; 13, Sussex West, divided from East Sussex by the high-road from Brighton to Cuckfield, thence through Crawley to the Surrey border; 14, Sussex East; 15, Kent East, divided from W. Kent by the Medway and its tributaries from its mouth nearly up to Staplehurst, thence the line is along the high-road through Cranbrooke to the Sussex border near Hawkeshurst; 16, Kent West; 17, Surrey; 18, Essex South, divided from North Essex by the highroad from Waltham and Epping to Chelmsford, thence along the river Blackwater to its mouth; 19, Essex North; 20, Hertfordshire; 21 Middlesex; 22, Berkshire; 23, Oxfordshire; 24, Buckinghamshire; 25, Suffolk East, divided from West Suffolk by the parallel of longitude 1° East from the meridian of Greenwich; 26, Suffolk West: (the detached portion upon which Newmarket stands is included with Cambridgeshire); 27, Norfolk East, divided from West Norfolk by the 1° East parallel of longitude; 28, Norfolk West; 29, Cambridgeshire, including the Newmarket detached portion of Suffolk; 30, Bedfordshire, including a detached portion of Huntingdonshire; 31, Huntingdonshire; 32, Northamptonshire; 33, Gloucestershire East, separated from West Gloucestershire by the Thames and Severn Canal, thence by the River Severn from the point of confluence of the canal up to Tewkesbury, and includes five detached portions of Worcestershire and one of Warwickshire; 34, Gloucesterhire West; 35, Monmouthshire (a detached portion is included in Herefordshire); 36, Herefordshire, including detached portions of Monmouthshire and Worcestershire (detached portions of Herefordshire are on the other hand included in Brecknockshire, Radnorshire, Shropshire, and Worcestershire); 37, Worcestershire, including detached portions of Herefordshire, Shropshire, Staffordshire, and Warwickshire (detached portions of this county are, on the other hand, included with Gloucestershire, Herefordshire, and Staffordshire); 38, Warwickshire (detached portions are included in Gloucestershire and Worcestershire); 39, Staffordshire, including a detached portion of Worcestershire, which in its turn includes an outlier of Staffordshire; 40, Shropshire, including an outlier of Herefordshire (a detached portion of Shropshire is included with Worcestershire; 41, Glamorganshire; 42, Brecknockshire, including a detached portion of Herefordshire; 43, Radnorshire, including a detached portion of Herefordshire; 44, Carmarthenshire; 45, Pembrokeshire; 46, Cardiganshire; 47, Montgomeryshire; 48, Merionethshire; 49, Carnarvonshire (the Llandudno peninsula and other portions of this county lying East of the river Conway are included with Denbighshire); 50, Denbighshire, including outliers of the county last named, also the detached portion of Flintshire; 51, Flintshire (the detached—or Overton—portion is included with Denbighshire); 52, Anglesey; 53, Lincolnshire South, divided from North Lincoln by the Witham from its mouth at Boston to Lincoln, thence by the Foss Dyke to the border of Nottinghamshire; 54, Lincolnshire North; 55, Leicester and Rutland, including a detached portion of Derbyshire; 56, Nottinghamshire; 57, Derbyshire (a detached portion is included with Leicester and Rutland); 58, Cheshire; 59, Lancashire South, separated from West Lancashire by the River Ribble; 60, Lancashire West (the Furness district of North Lancashire is included with Westmorland); 61, Yorkshire South-East, being the East Riding of Yorkshire; 62, Yorkshire North-East, divided from N. W. Yorkshire by the Rivers Wiske and Swale; 63, Yorkshire South-West, equivalent to the Southern division of the West Riding, divided from Mid-West Yorkshire by the Leeds and Liverpool Canal and by the River Aire below Leeds; 64, Yorkshire Mid-West, equivalent to the Northern portion of the West Riding minus the Dent and Sedbergh district, which is included with North-West Yorkshire; 65, Yorkshire North-West, equivalent to the western half of the North Riding with the addition of the Dent and Sedbergh district of the West Riding; 66, Durham (the detached portions are included with Northumberland and Cheviotland); 67, Northumberland South, includes a detached portion of Durham, and is separated from Cheviotland by the River Coquet and a line continued to Carter Fell from the Linn bridge; 68, Cheviotland, equivalent to North Northumberland plus a detached portion of Durham; 69, Westmorland and Furness; 70, Cumberland; 71, Isle of Man; 72, Dumfriesshire; 73, Kirkcudbrightshire; 74, Wigtonshire; 75, Ayrshire; 76, Renfrewshire; 77, Lanarkshire; 78, Peeblesshire; 79, Selkirkshire; 80, Roxburghshire; 81, Berwickshire; 82, Haddingtonshire; 83, Edinburghshire; 84, Linlithgowshire; 85, Fife and Kinross; 86, Stirlingshire, including the detached portion of Dumbartonshire; 87, Perth West and Clackmannan, separated from Mid-Perth by the line of watershed which divides tributaries of the Tay from those of the Forth; 88, Perth Mid, separated from East Perth by the rivers Garry and Tay; 89, Perth East; 90, Forfarshire; 91, Kincardineshire; 92, Aberdeenshire South, separated from N. Aberdeen by the watershed line which runs E. and W. from Inverury; 93, Aberdeenshire North; 94, Banffshire; 95, Elginshire, including the detached portion of Inverness-shire which separates the two portions of Elginshire; 6, Easterness (to form the vice-counties "Easterness" and "Westerness," Invernessshire is first divided by the line of watershed between the East and West of Scotland, continued along Loch Errickt to the Perthshire border, the Eastern portion—with Nairnshire added—being called Easterness, and the Western portion—with the detached portion of Argyleshire situated N. W. of Loch Linnhe—is called Westerness); 97, Westerness (see 96 for definition); 98, Main Argyleshire, is what is left of the county after the separation of Cantire, the Islands, and the portion N. W. of Loch Linnhe; 99, Dumbartonshire (the detached portion is included with Stirlingshire); 100, Clyde Islands, Bute, Arran, Cumbray, and Ailsa Craig; 101, Cantire, this peninsula is separated from Argyleshire by the Crinan Canal; 102, Ebudes South, the Islands of Jura, Colonsay, and Islay; 103, Ebudes Mid, the Islands of Mull, Coll, Tirree, Staffa, Iona, etc.; 104, Ebudes North, the Islands of Skye, Canna, Rum, Muck, Eig, etc.; 105, Ross-shire West, separated from East Ross-shire by the watershed parting East and West Scotland, and including some of the fragments of Cromarty county; 106, Ross-shire East, including fragments of Cromarty and Nairnshire; 107, Sutherland South-East, divided from North-West Sutherland by the watershed line parting the East and West sides of Scotland; 108, Sutherland North-West; 109, Caithness; 110, Hebrides; 111, Orkneys; 112, Shetland Islands; 113, Londonderry County (the city of Londonderry is included with Donegal county); 114, Antrim County; 115, Down County; 116, Armagh County; 117, Monaghan County; 118, Tyrone County; 119, Donegal County, including Londonderry City; 120, Fermanagh County; 121, Cavan County; 122, Louth County; 123, Meath County; 124, Dublin County; 125, Kildare County; 126, Wicklow County; 127, Wexford County; 128, Carlow County; 129, Kilkenny County; 130, Queen's County; 131, King's County; 132,

Westmeath County; 133, Longford County; 134, Roscommon County; 135, Leitrim County; 136, Sligo County; 137, Mayo East, separated from West Mayo by the railway from Ballina to the head of Lough Mask; 138, Mayo West; 139, Galway West, separated from East Galway by Lough Corrib; 140, Galway East; 141, Clare County; 142, Limerick County; 143, Tipperary North, divided from South Tipperary by the line of watershed; 144, Tipperary South; 145, Waterford County; 146, Cork North, divided from South Cork by the River Lee; 147, Cork South; 148, Kerry County.

www.ingramcontent.com/pod-product-compliance
Lightning Source LLC
Chambersburg PA
CBHW030906170426
43193CB00009BA/752